CHEMICAL EQUILIBRIA IN SOLUTION

Dependence of Rate and Equilibrium Constants on Temperature and Pressure

Ellis Horwood and Prentice Hall
are pleased to announce their collaboration in a new imprint whose list will encompass outstanding works by world-class chemists, aimed at professionals in research, industry and academia. It is intended that the list will become a by-word for quality, and the range of disciplines in chemical science to be covered are:

ANALYTICAL CHEMISTRY
ORGANIC CHEMISTRY
INORGANIC CHEMISTRY
PHYSICAL CHEMISTRY
POLYMER SCIENCE & TECHNOLOGY
ENVIRONMENTAL CHEMISTRY
CHEMICAL COMPUTING & INFORMATION SYSTEMS
BIOCHEMISTRY
BIOTECHNOLOGY

Ellis Horwood PTR Prentice Hall
PHYSICAL CHEMISTRY SERIES

Series Editors:
Ellis Horwood, M.B.E.
Professor T J Kemp, University of Warwick

Current Ellis Horwood PTR Prentice Hall
Physical Chemistry Series titles

Blandamer	**CHEMICAL EQUILIBRIA IN SOLUTION: Dependence of Rate and Equilibrium Constants on Temperature and Pressure**
Bugayenko	**HIGH ENERGY CHEMISTRY**
Horspool & Armesto	**ORGANIC PHOTOCHEMISTRY: A Comprehensive Treatment**
Navratil	**NUCLEAR CHEMISTRY**

CHEMICAL EQUILIBRIA IN SOLUTION

Dependence of Rate and Equilibrium Constants on Temperature and Pressure

MICHAEL J. BLANDAMER, B.Sc., Ph.D., D.Sc.
Professor of Physical Chemistry, University of Leicester

ELLIS HORWOOD PTR PRENTICE HALL
NEW YORK LONDON TORONTO SYDNEY TOKYO SINGAPORE

First published in 1992 by
ELLIS HORWOOD LIMITED
Market Cross House, Cooper Street,
Chichester, West Sussex, PO19 1EB, England

A division of
Simon & Schuster International Group
A Paramount Communications Company

British Library Cataloguing in Publication Data

A catalogue record for this book is available from the British Library

0–13–131731–8

Library of Congress Cataloging-in-Publication Data

available from the publishers

Printed and bound in Great Britain
by Bookcraft Limited, Midsomer Norton, Avon

Table of contents

Preface vii
Units and symbols viii
1 Affinity for chemical reaction 1
2 Composition of solutions; chemical potentials 19
3 Chemical equilibria in solution 47
4 Dependence of equilibrium and rate constants on pressure 59
5 Dependence of equilibrium and rate constants on temperature —linear equations 85
6 Dependence of equilibrium and rate constants on temperature —extended equations 122
7 Dependence of equilibrium composition and rate constants on temperature and pressure 136
Index 143

To Anne

Preface

A chemical equilibrium in a given closed system at constant temperature and pressure is at a minimum in Gibbs energy. The chemical composition at equilibrium is characterized by various thermodynamic parameters including a chemical equilibrium constant. When either temperature or pressure is changed, these properties change, yielding further information about the chemical equilibrium. This text explores how these dependences are treated and how information is extracted from experimental data. Related treatments are discussed, describing the dependences on temperature and pressure of rate constants. Examples are taken from the chemical literature although no attempt is made to survey the vast amount of material describing chemical equilibria and reactions in solutions. Nevertheless the hope is that this book will help chemists who wish to contribute to this fascinating subject.

I acknowledge at the outset a considerable debt to many colleagues and friends for their encouragement and advice. The list is long and includes John Burgess, Jan B. F. N. Engberts, Ross E. Robertson, John M. W. Scott and Martyn C. R. Symons.

Mike Blandamer

Units and symbols

For the most part we use the units, symbols and names of variables as recommended by IUPAC. We also pay attention to the form of equations, ensuring that they are dimensionally correct. There remains however the old 'chestnut' of what to do with operations which require the logarithm of a variable which is not dimensionless. For example many equations involve the expression $\ln T$ where T is the temperature. To avoid the necessity of hunting for the term in ln(unit) we write $\ln(T/\mathrm{K})$. We have used K^0 to indicate standard equilibrium constants at the standard pressure p^0. Thus $K(T;p)$ is an equilibrium constant at pressure p and temperature T. $K^0(\mathrm{T})$ is a standard equilibrium constant at temperature T. In other words $K^0(T)$ is not dependent on pressure, whereas $K(T;p)$ is.

1

Affinity for chemical reaction

1.1 INTRODUCTION

For a closed system comprising one phase, a set of independent variables defines all other properties, the dependent variables [1]. For the most part we will use the independent variables, T; p; n_1; n_2; n_3; ...; n_j for chemical substances 1, 2, 3, 4, ..., j. If n_i describes the number of moles of all chemical substances in a given system, the dependent variable called volume, V, is defined by the equation, $V = V[T; p; n_i]$. The property V of a system is a state property (state variable). If chemical reaction produces a change of state, the change in state variable X, ΔX is independent of the path between the two states and the rate of change; $\Delta X = X(\text{state II}) - X(\text{state I})$. Among the important state variables which we consider here are the thermodynamic energy U, the entropy S, the enthalpy H and the Gibbs energy G of a system.

The chemical composition of a system can change through both forward and reverse processes in a *chemically reversible* reaction. The term *thermodynamically reversible* identifies a change under equilibrium conditions. In other words, the term 'reversible' is used in two ways [2].

The term 'irreversible' also has two meanings. A given chemical reaction involves solute X in an aqueous solution where the chemical reaction is, $X(\text{aq}) \rightarrow Y(\text{aq})$. Suppose we monitor property P of the system as a function of time to chart the course of reaction. When no further change in P occurs and if no chemical substance X remains in the system, the reaction is chemically irreversible. The term thermodynamically irreversible means a change under non-equilibrium conditions.

References to section 1.1

[1] E. A. Guggenheim. *Thermodynamics*, North Holland, 1950, 2nd. edn.
[2] R. K. Boyd. *Chem. Rev.*, 1977, **77**, 93.

1.2 LAWS OF THERMODYNAMICS

The energy, U, of a closed system increases when work is done by the surroundings on the system and heat Q flows into the system from the surroundings. For the most part we are concerned with 'p–V' work [1]. A given closed system at temperature T

Table 1.2.1. The Second Law of Thermodynamics

LAW: $\mathrm{d}S = (Q/T) + \mathrm{d_i}S$

$\mathrm{d_i}S = 0$: reversible process, equilibrium transformation.
$\mathrm{d_i}S > 0$: irreversible process; spontaneous process; chemical reaction; natural process.
$\mathrm{d_i}S < 0$: forbidden

Table 1.2.2. Chemical reaction

Closed system; fixed T and p

$$w_j = w_j(t = 0) + v_j M_j \xi;$$

where $v_j = [1]$; $M_j = [\mathrm{kg\,mol^{-1}}]$; $\xi = [\mathrm{mol}]$

$$\mathrm{d}n_j = v_j\,\mathrm{d}\xi \qquad \text{(a)}$$

Rate of reaction, $J = \mathrm{d}\xi/\mathrm{d}t = (1/v_j)(\mathrm{d}n_j/\mathrm{d}t)$

$$\mathrm{d}c_j/\mathrm{d}t = V^{-1}\,\mathrm{d}n_j/\mathrm{d}t = v_j J/V$$

A mnemonic: v_j is **P**(ositive) for **P**(roducts).

is held at volume V by applied pressure p. If the volume changes from V to $(V + \mathrm{d}V)$, and if at all stages the pressure p of the system equals the external pressure, $\mathrm{d}U = Q - p\,\mathrm{d}V$; the first law of thermodynamics.

Two possible modes of transformation between states I and II of a system are (i) reversible (equilibrium) and (ii) irreversible (spontaneous) (Table 1.2.1). If the change at temperature T is thermodynamically reversible, $Q = T\,\mathrm{d}S$. The variables S and T are extensive and intensive respectively. For a spontaneous process, $Q < T\,\mathrm{d}S$. Three important features of spontaneous changes are (i) the extent of change, (ii) the affinity for change, and (iii) the rate of change.

Within a given closed reaction vessel [2, 3], chemical substance j is involved in chemical reaction. At time zero, the mass of substance j equals $w_j(t = 0)$. Mass w_j at time t is related [4, 5] to $\mathrm{d}\xi$, the extent of reaction and the stoichiometry v_j (Table 1.2.2). But $w_j/M_j = n_j$, the amount of substance j where $\mathrm{d}n_j$ is given by equation (a) of Table 1.2.2. Rate of reaction J at time t is given by the differential of equation (a) with respect to time (Table 1.2.2) and is therefore related to the rate of change of n_j. Both ξ and J are extensive properties [6].

References to section 1.2

[1] D. Kivelson; I. Oppenhein. *J. Chem. Educ.*, 1966, **43**, 233.
[2] T. M. Barbaro; P. L. Corio. *J. Chem. Educ.*, 1980, **57**, 243.
[3] P. L. Corio. *J. Phys. Chem.*, 1984, **88**, 1825.
[4] K. J. Laidler; N. Kallay. *Kem. Ind.*, 1988, **37**, 183.

[5] P. G. Wright. *Educ. in Chem.*, 1986, 96.
[6] For an interesting discussion see O. Redlich. *J. Chem. Educ.*, 1970, **47**, 154.

1.3 AFFINITY FOR SPONTANEOUS CHEMICAL REACTION

The extent of reaction $d\xi$ accompanying spontaneous chemical reaction in a closed system at temperature T is related to $d_i S$ by equation (a) of Table 1.3.1. The change in entropy dS is given by equation (b) of Table 1.3.1. The inequality $A\,d\xi \geqslant 0$ refers to spontaneous reaction where the affinity for spontaneous change is A and extent of reaction is $d\xi$. Because dt is positive, both $A\,d\xi$ and the product of affinity and velocity, AJ are non-negative (Table 1.3.2).

A closed system in which both the velocity of change and the affinity for change are zero is in thermodynamic equilibrium with its surroundings (Table 1.3.2). This condition forms the basis of this text. For a system undergoing spontaneous chemical reaction, the change in composition is represented by the symbol $d\xi$, the chemical reaction being driven by the affinity for spontaneous change A. Eventually (i.e. limit$(t \to \infty)$) no further reaction occurs, at which stage the solution is in equilibrium with the surroundings, the affinity for spontaneous change being zero.

The change in thermodynamic energy, dU, at temperature T and pressure p is related (Table 1.3.3) to the change in entropy and volume, together with the extent of reaction $d\xi$; the equation for dU is an axiom in thermodynamics [1]. The thermodynamic energy U, the dependent variable, is expressed (Table 1.3.4) in terms of three independent variables S, V and ξ, where the composition of a closed system is described by the symbol ξ. Therefore instead of writing $X = X(S; V; n_i)$, we write $X = X(S; V; \xi)$ for the thermodynamic variable X.

Table 1.3.1. Affinity and change

Definition*: $T\,d_i S = A\,d\xi \geqslant 0$ (a)

But (Table 1.2.1) $dS = (Q/T) + d_i S$

Therefore, $dS = (Q/T) + (A/T)\,d\xi$ (b)

*Units: $[K][J\,K^{-1}] = [J\,mol^{-1}][mol]$

Table 1.3.2. Spontaneous chemical reaction

Second Law: $A\,d\xi \geqslant 0$; $A(d\xi/dt) = AJ \geqslant 0$

Condition I
Spontaneous chemical reactions: A and J have the same sign.

Condition II
Chemical equilibria: Both A and J are zero.

Table 1.3.3. Thermodynamic energy—general equation

$A\,d\xi > 0$; spontaneous change. $A\,d\xi = 0$; equilibrium.
Axiom*, $dU = T dS - p dV - A\,d\xi$
Changes in (a) extensive properties, U, S, V, and ξ, and
(b) intensive properties, T, p and A

*Units: $[J] = [K][J\,K^{-1}] - [N\,m^{-2}][m^3] - [J\,mol^{-1}][mol]$

Table 1.3.4. Thermodynamic energy and change; $U = U[S; V; \xi]$

Dependent variable, U; independent variables, S, V and ξ.
General differential

$$dU = (\partial U/\partial S)_{V,\xi}\,dS + (\partial U/\partial V)_{S;\xi}\,dV + (\partial U/\partial \xi)_{S;V}\,d\xi$$

Hence* (Table 1.3.3), (a) $T = (\partial U/\partial S)_{V;\xi}$,
(b) $p = -(\partial U/\partial V)_{S;\xi}$,
and (c) $A = -(\partial U/\partial \xi)_{S;V}$.

*Units: (a) $[K] = [J][J\,K^{-1}]$; (b) $[N\,m^{-2}] = [J]/[m^3]$; (c) $[J\,mol^{-1}] = [J]/[mol]$

The coefficients of dS, dV and $d\xi$ in the complete differential of U (Table 1.3.4) are compared with the corresponding coefficients of dU in Table 1.3.3. Three equations are obtained for the intensive properties T, p and affinity A.

We now suppose that a solution is held within a vessel which is linked directly to a piston and cylinder arrangement. Furthermore we use the piston to hold the volume of the solution constant throughout the course of the reaction. In other words, V = constant and dV is zero. Another condition we impress on the chemical reaction requires that the entropy S is constant; i.e. dS = zero. It is not immediately obvious how one might satisfy this criterion from a practical standpoint. Granted for the moment that the two conditions, constant V and constant S, can be satisfied an important result follows from the equations in Table 1.3.3. Thus, $dU = -A\,d\xi$. For a spontaneous process, $A\,d\xi > 0$; hence $dU < 0$. Chemical reaction at constant S and V lowers the thermodynamic energy; U is therefore the thermodynamic potential function for chemical reaction at constant S and V. At equilibrium, the affinity A is zero and so, in a plot of U against ξ, thermodynamic equilibrium corresponds to a minimum in U. This conclusion is independent of the nature and rate of processes taking place in the reaction vessel. But we are out of luck because it is not obvious how one sets up such a system in the laboratory. We turn the problem on its head and ask: what conditions can be satisfied in the laboratory with modest capital investment? The answer is: constant temperature and pressure. Therefore we seek an extensive function of state for closed systems where (i) the independent variables are T, p and ξ, and (ii) the new function provides information concerning the direction

Table 1.3.5. Gibbs energy

$$G = U + pV - TS$$

Differential. $\mathrm{d}G = \mathrm{d}U + p\,\mathrm{d}V + V\mathrm{d}p - T\mathrm{d}S - S\mathrm{d}T$
But (Table 1.3.3) $\mathrm{d}U = T\mathrm{d}S - p\,\mathrm{d}V - A\,\mathrm{d}\xi$
Therefore*, $\mathrm{d}G = -S\mathrm{d}T + V\mathrm{d}p - A\,\mathrm{d}\xi$ with, $A\,\mathrm{d}\xi > 0$

*Units: $[\mathrm{J}] = [\mathrm{J\,K^{-1}}][\mathrm{K}] + [\mathrm{m^3}][\mathrm{N\,m^{-2}}] - [\mathrm{J\,mol^{-1}}][\mathrm{mol}]$

Table 1.3.6. Gibbs energy and affinity

$$G = G[T; p; \xi]$$

Dependent variable, G; independent variables: T, p, and ξ.
General differential.

$$\mathrm{d}G = (\partial G/\partial T)_{p;\xi}\,\mathrm{d}T + (\partial G/\partial p)_{T;\xi}\,\mathrm{d}p + (\partial G/\partial \xi)_{T;p}\,\mathrm{d}\xi$$

Therefore* (with Table 1.3.5),
(a) $S = -(\partial G/\partial T)_{p;\xi}$ (b) $V = (\partial G/\partial p)_{T;\xi}$ (c) $A = -(\partial G/\partial \xi)_{T;p}$

*Units: (a) $[\mathrm{J\,K^{-1}}] = [\mathrm{J}]/[\mathrm{K}]$; (b) $[\mathrm{m^3}] = [\mathrm{J}]/[\mathrm{N\,m^{-2}}]$; (c) $[\mathrm{J\,mol^{-1}}] = [\mathrm{J}]/[\mathrm{mol}]$

of spontaneous change. The required property is the Gibbs energy, G, defined (Table 1.3.5) in terms of U together with the products, pV and TS.

The equations defining G and $\mathrm{d}U$ (Tables 1.3.5 and 1.3.3 respectively) provide an equation for $\mathrm{d}G$ in terms of changes in temperature, and pressure. For a closed system, the extensive function of state G is defined by the independent variables, T, p, and ξ (Table 1.3.6). Comparison of the equation for $\mathrm{d}G$ obtained as the general differential and the equation for $\mathrm{d}G$ (Table 1.3.5) leads to equations for V, S, and A in terms of partial differentials of G. Relationships between these quantities [2–4] are summarized in Fig. 1.3.1. The figure highlights the central role of the Gibbs energy in our discussions. The Roman numeral I signals the thermodynamic variables which are obtained by one differential operation on G. Roman numerals II and III signal properties which require one and two more differential operations respectively. An interesting partial derivative is $(\partial G/\partial p)_{T;p}$. Although we cannot measure G for a given system, its derivative with respect to pressure links directly to a quantity which is readily measured, the volume. The partial differential $-(\partial G/\partial T)_{P;\xi}$ equals the entropy S. The partial differential $-(\partial G/\partial \xi)_{T;p}$, the ratio of differential changes in two extensive variables, equals the affinity A, an intensive property. At fixed temperature and pressure, $\mathrm{d}G = -A\,\mathrm{d}\xi$. But $A\,\mathrm{d}\xi > 0$ for all spontaneous processes. Hence for a closed system at fixed T and fixed p, $\mathrm{d}G < 0$; all spontaneous processes reduce the Gibbs energy of a system. At equilibrium, $(A^{\mathrm{eq}})_{T;p} = -(\partial G/\partial \xi)^{\mathrm{eq}}_{T;p} = 0$. Hence G^{eq} is a minimum. Therefore G is the thermodynamic potential function for

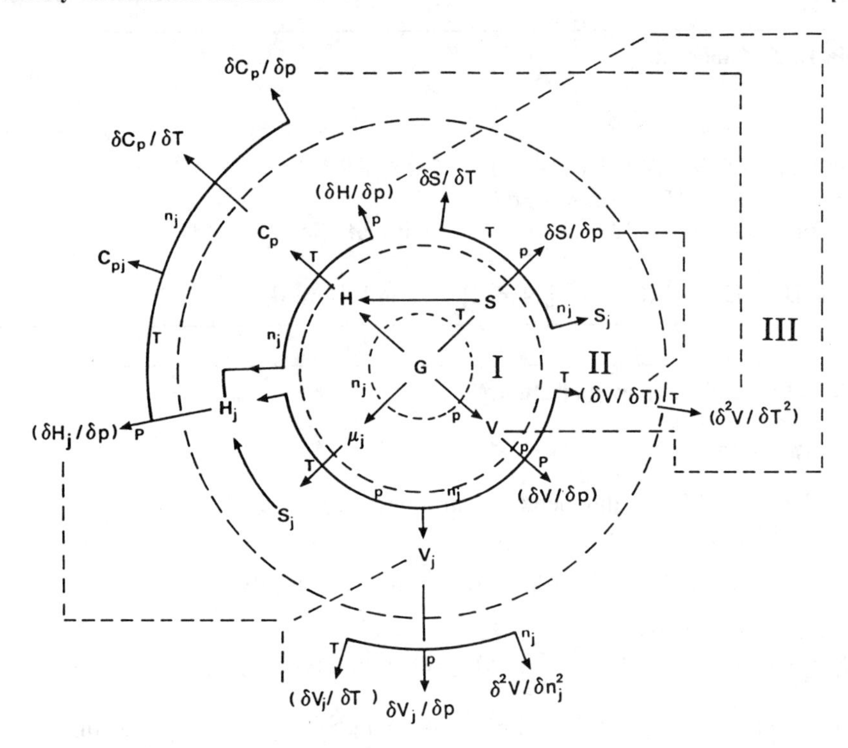

Fig. 1.3.1. Gibbs energy and related properties.

processes at fixed T and fixed p. We are on familiar ground here and armed with a very powerful weapon. For systems at fixed temperature and fixed pressure, we have a set of criteria for spontaneous changes and thermodynamic equilibrium. Nevertheless, we had to sacrifice something along the way. The Gibbs energy G is a contrived property; the thermodynamic energy U is hidden within it. The condition for equilibrium at fixed temperature and pressure is a minimum in G and not, as we might have hoped, a minimum in thermodynamic energy. We return to the definition of G in terms of independent variables, T, p and n_i: $G = G[T; p; n_i]$. This equation applies to both equilibrium and non-equilibrium states. For a given closed system, the equilibrium state is uniquely defined [5] by G^{eq} and n_i^{eq} at constant T and p; $G^{eq} = G[T; p; n_i^{eq}]$.

References to section 1.3

[1] M. L. McGlashan. *Chemical Thermodynamics*, Academic Press, London, 1979.
[2] M. J. Blandamer; J. Burgess. *Education in Chemistry*, 1987, 85.
[3] M. Isihara. *Bull. Chem. Soc. Japan*, 1986, **59**, 2932.
[4] R. Gilmore. *J. Chem. Phys.*, 1982, **77**, 5853; 1981, **75**, 5964.
[5] F. van Zeggeren; S. H. Storey. *The Computation of Chemical Equilibria*, Cambridge University Press, 1970.

1.4 ENTHALPIES AND ISOBARIC HEAT CAPACITIES

An important extensive state variable is the enthalpy H (Table 1.4.1). The enthalpy of a system is important to our discussion of the dependence of G on temperature. We direct our attention to systems held at constant temperature and pressure for which the key quantity is the Gibbs energy G (Fig. 1.4.1). In other words we define the state of a closed system using the independent variables T, p and ξ. The general differential of the equation $H = H[T; p; \xi]$ leads to an equation for $\mathrm{d}H$ in terms of the changes, $\mathrm{d}T$, $\mathrm{d}p$ and $\mathrm{d}\xi$ (Table 1.4.2). Combination of the equation defining H and the First Law provides an important equation for the heat Q in terms of changes in temperature, pressure and composition. For the most part our interest centres on the properties of systems at constant pressure. Under these conditions heat Q is given by the sum of two terms:

$$Q = (\partial H/\partial T)_{p;\xi}\,\mathrm{d}T + (\partial H/\partial \xi)_{T;p}\,\mathrm{d}\xi \qquad (1.4.1)$$

Each term is a product of intensive and extensive properties. The first term on the right-hand side of equation (1.4.1) describes the associated change in temperature and the (extensive) change in H at constant composition and pressure. The second term describes the (extensive) change in composition and the (intensive) dependence of H on ξ at fixed temperature and pressure. The impact of heat Q passing into a system at fixed pressure is related [1] to changes in temperature and composition. The partial derivative $(\partial H/\partial T)_{p;\xi}$ is the isobaric heat capacity at constant composition, $C_\mathrm{p}(\xi)$ (Table 1.4.3); the composition is 'frozen' with $\mathrm{d}\xi = 0$. Isobaric heat capacity C_p is an

---

Table 1.4.1. Enthalpy

Definition: $H = U + pV$
Hence, $\mathrm{d}H = \mathrm{d}U + p\,\mathrm{d}V + V\,\mathrm{d}p$.
But (Table 1.3.3) $\mathrm{d}U = T\,\mathrm{d}S - p\,\mathrm{d}V - A\,\mathrm{d}\xi$
Then, $\mathrm{d}H = T\,\mathrm{d}S + V\,\mathrm{d}p - A\,\mathrm{d}\xi$
If $G = U + pV - TS$ and $H = U + pV$ then $G = H - TS$

Table 1.4.2. Thermodynamic function, H

$H = H[T; p; \xi]$
Dependent variable. H. Independent variables: T, p and ξ.
Hence, $\mathrm{d}H = (\partial H/\partial T)_{p;\xi}\,\mathrm{d}T + (\partial H/\partial p)_{T;\xi}\,\mathrm{d}p + (\partial H/\partial \xi)_{T;p}\,\mathrm{d}\xi$
But (Table 1.4.1), $\mathrm{d}U = \mathrm{d}H - p\,\mathrm{d}V - V\,\mathrm{d}p$

$$\mathrm{d}U = (\partial H/\partial T)_{p;\xi}\,\mathrm{d}T + [(\partial H/\partial p)_{T;\xi} - V)\,\mathrm{d}p - p\,\mathrm{d}V + (\partial H/\partial \xi)_{T;p}\,\mathrm{d}\xi$$

According to the First Law, $Q = \mathrm{d}U + p\,\mathrm{d}V$
Hence, $Q = (\partial H/\partial T)_{p;\xi}\,\mathrm{d}T + [(\partial H/\partial p)_{T;\xi} - V]\,\mathrm{d}p + (\partial H/\partial \xi)_{T;p}\,\mathrm{d}\xi$
or $Q = C_\mathrm{p}(\xi)\,\mathrm{d}T + [(\partial H/\partial p)_{T;\xi} - V]\,\mathrm{d}p + (\partial H/\partial \xi)_{T;p}\,\mathrm{d}\xi$

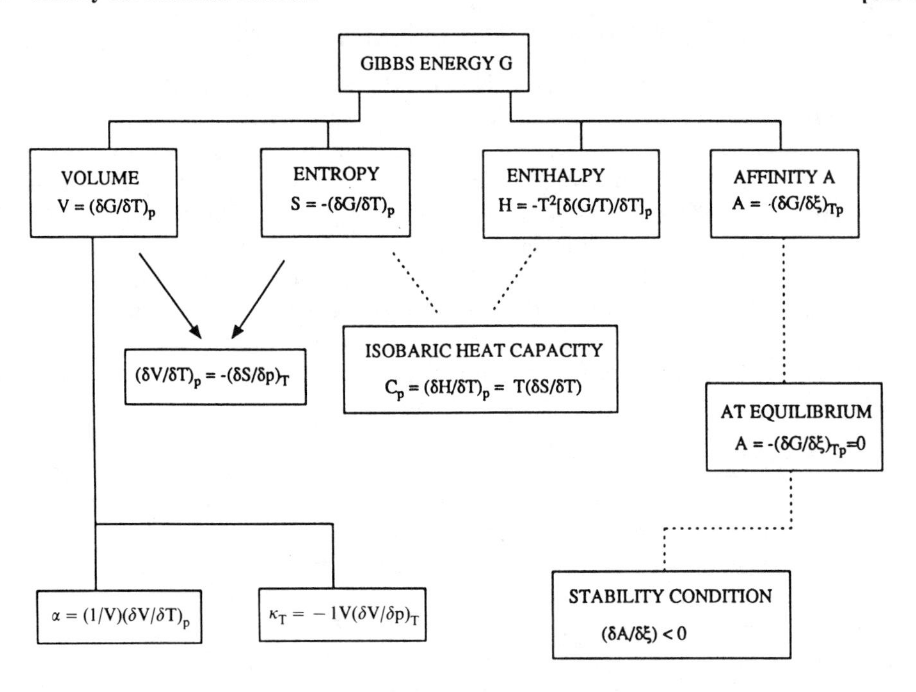

Fig. 1.4.1. Gibbs energy and related properties.

Table 1.4.3. Heat capacity, C

*Heat capacity $C = Q/\mathrm{d}T$
At constant pressure, $C_\mathrm{p} = (\partial H/\partial T)_p$
At constant pressure and composition, $C_\mathrm{p}(\xi) = (\partial H/\partial T)_{p;\xi}$
At constant volume, $C_\mathrm{v} = (\partial U/\partial T)_V$
At constant volume and composition, $C_\mathrm{v}(\xi) = (\partial U/\partial T)_V$

*Units: $C = [\mathrm{J\,K^{-1}}]$

---

extensive property of a system. $(\partial H/\partial \xi)_{T;p}$, the differential change in enthalpy with respect to the extent of reaction, measures the change in H when ξ advances by unity.

In addition to $C_\mathrm{p}(\xi)$, two factors control isobaric heat capacities: (a) the dependence of H on ξ, and (b) the dependence of ξ on T. A large C_p is favoured [2] by marked sensitivities of H on ξ and of ξ on T.

$$\text{Isobaric heat capacity } C_\mathrm{p} = C_\mathrm{p}(\xi) + (\partial H/\partial \xi)_{T;p}(\partial \xi/\partial T)_p \tag{1.4.2}$$

Another important thermodynamic variable is the Helmholtz energy, F (Table 1.4.4). At constant T and constant V, $\mathrm{d}F = -A\,\mathrm{d}\xi$. But $A\,\mathrm{d}\xi > 0$. Hence for all spontaneous processes (e.g. chemical reaction) at constant T and constant V, $\mathrm{d}F < 0$. At equilibrium, F is a minimum; $(\partial F/\partial \xi)^{\mathrm{eq}}_{T;V} = 0$.

Table 1.4.4. Helmholtz energy, F

Definition: $F = U - TS$
Then, $dF = dU - T\,dS - S\,dT$.
But (Table 1.3.3), $dU = T\,dS - p\,dV - A\,d\xi$
Therefore, $dF = -S\,dT - p\,dV - A\,d\xi$
Independent variables; T, V and ξ. Dependent variable, F

$$F = F[T; V; \xi]$$

$$dF = (\partial F/\partial T)_{V;\xi}\,dT + (\partial F/\partial V)_{T;\xi}\,dV + (\partial F/\partial \xi)_{T;V}\,d\xi$$

Therefore, $S = -(\partial F/\partial T)_{V;\xi}$
and* $p = -(\partial F/\partial V)_{T;\xi}$ and, $A = -(\partial F/\partial \xi)_{T;V}$

*Units: $[\mathrm{N\,m^{-2}}] = [\mathrm{J}]/[\mathrm{m^3}]$

Thermodynamic variables G and F are related; $G = F + pV$. Properties F and G are practical thermodynamic potential functions. We can set up experiments in the laboratory under conditions which conform to the requirements for either to act as the thermodynamic potential function.

References to section 1.4

[1] M. J. Blandamer; J. Burgess; J. M. W. Scott. *Ann. Rept. Progr. Chem., Sect. C*, 1985, **82**, 77.
[2] D. H. Everett. *Disc. Faraday Soc.*, 1957, **24**, 216.

1.5 LINKING EQUATIONS

Many important equations are derived by following the procedures set out in Table 1.5.1. Examples (a) and (b) recognize that G is a thermodynamic function of state. This type of manipulation leads to 10^{10} Maxwell's equations [1, 2].

As a basis for discussion in later chapters we need to develop equations which describe the dependences on T and p of thermodynamic variables G, H, V and S (Tables 1.5.2 to 1.5.3). For systems at fixed ξ, the dependence of G/T on T at fixed pressure and related to the enthalpy H (Table 1.5.2). We assume that G is a continuous function of T, p and composition. The condition 'fixed ξ' means that H characterizes the dependence of (G/T) on T at fixed composition. The dependence of H on T at constant composition and pressure is characterized by the isobaric heat capacity, $C_p(\xi)$.

The Gibbs–Helmholtz equation prompts consideration of the dependent variable (G/T) as a function of the independent variables T, p and ξ (Table 1.5.3). Next we consider the dependent variable V, the volume (Table 1.5.4). The partial differential of V with respect to T at fixed pressure and composition defines $\alpha(\xi)$; the related

Table 1.5.1. Maxwell's equations

(a) From Table 1.3.6, $V = (\partial G/\partial p)_{T;\xi}$ and, $S = -(\partial G/\partial T)_{p;\xi}$
But G is a function of state.
Hence $(\partial^2 G/\partial p \partial T)_\xi = (\partial^2 G/\partial T \partial p)_\xi$, or $(\mathrm{d}/\mathrm{d}T)(\partial G/\partial p) = (\mathrm{d}/\mathrm{d}p)(\partial G/\partial T)$
Therefore $(\partial V/\partial T)_{p;\xi} = -(\partial S/\partial p)_{T;\xi}$
Similarly, $(\partial V/\partial T)_{p;A} = -(\partial S/\partial p)_{T;A}$
(b) From $\mathrm{d}^2 G/\mathrm{d}\xi\,\mathrm{d}p = \mathrm{d}^2 G/\mathrm{d}p\,\mathrm{d}\xi$, or $(\mathrm{d}/\mathrm{d}\xi)(\partial G/\partial p) = (\mathrm{d}/\mathrm{d}p)(\partial G/\partial \xi)$
Hence $(\partial V/\partial \xi)_{T;p} = -(\partial A/\partial p)_{T;\xi}$
(c) From $\mathrm{d}^2 H/\mathrm{d}\xi\,\mathrm{d}T = \mathrm{d}^2 H/\mathrm{d}T\,\mathrm{d}\xi$ (see Table 1.4.3)
Hence, $(\partial C_\mathrm{p}/\partial \xi)_{T;p} = (\mathrm{d}/\mathrm{d}T)(\partial H/\partial \xi)_{T;p}$
(d) From $\partial^2 F/\partial T \partial V = \partial^2 F/\partial V \partial T$. (see Table 1.4.4)
Then, $(\partial/\partial T)(\partial F/\partial V) = (\partial/\partial V)(\partial F/\partial T)$
Hence $(\partial p/\partial T)_{V;\xi} = (\partial S/\partial V)_{T;\xi}$

Table 1.5.2. Gibbs–Helmholtz equations

Gibbs function: $G = H - TS$. But (Table 1.3.6), $S = -(\partial G/\partial T)_{p;\xi}$
Then, $H = G - T(\partial G/\partial T)_{p;\xi}$ or $[\partial(G/T)/\partial T]_{p;\xi} = -H/T^2$
or $H = -T^2[\partial(G/T)/\partial T]_{p;\xi}$ and $H = [\partial(G/T)/\partial(1/T)]_{p;\xi}$

Table 1.5.3. Dependent variable (G/T)

$$(G/T) = (G/T)[T; p; \xi]$$

$\mathrm{d}(G/T) = [\partial(G/T)/\partial T]_{p;\xi}\,\mathrm{d}T + [\partial(G/T)/\partial p]_{T;\xi}\,\mathrm{d}p + [\partial(G/T)/\partial \xi]_{T;p}\,\mathrm{d}\xi$
Then (Tables 1.3.6 and 1.5.2),
$\mathrm{d}(G/T) = -(H/T^2)\,\mathrm{d}T + (V/T)\,\mathrm{d}p + (1/T)(\partial G/\partial \xi)_{T;p}\,\mathrm{d}\xi$

Table 1.5.4. Dependence of volume on temperature and pressure

$V = V[T; p; \xi]$

$$\mathrm{d}V = (\partial V/\partial T)_{p;\xi}\,\mathrm{d}T + (\partial V/\partial p)_{T;\xi}\,\mathrm{d}p + (\partial V/\partial \xi)_{T;p}\,\mathrm{d}\xi$$

Isobaric thermal expansivity at constant composition, $\alpha(\xi) = (1/V)(\partial V/\partial T)_{p;\xi}$
From Table 1.5.1 $(\partial S/\partial p)_{T;\xi} = -V\alpha(\xi)$
Isothermal compressibility at constant composition, $\kappa(\xi) = -(1/V)(\partial V/\partial p)_{T;\xi}$
Hence, $\mathrm{d}V = \alpha(\xi)V\,\mathrm{d}T - \kappa(\xi)V\,\mathrm{d}p + (\partial V/\partial \xi)_{T;p}\,\mathrm{d}\xi$

Table 1.5.5. Dependence of enthalpy and isobaric heat capacity on pressure

From Table 1.4.2, $H = H[T; p; \xi]$. From Table 1.4.1, $H = G + TS$
Then, $(\partial H/\partial p)_{T;\xi} = (\partial G/\partial p)_{T;\xi} + T(\partial S/\partial p)_{T;\xi}$
But (Table 1.3.6) $V = (\partial G/\partial p)_{T;\xi}$
and (Table 1.5.1) $(\partial S/\partial p)_{T;\xi} = -(\partial V/\partial T)_{p;\xi}$
Hence, $(\partial H/\partial p)_{T;\xi} = V - T(\partial V/\partial T)_{p;\xi}$
Therefore, $H(T; p_2; \xi) - H(T; p_1; \xi) = \int_{p_1}^{p_2} [V - T(\partial V/\partial T)_{p;\xi}]\,\mathrm{d}p$
Dependence of $C_p(\xi)$ on pressure
From above, $[\partial[\partial H/\partial p]/\partial T]_\xi = [\partial[V - T\{\partial V/\partial T\}_p]/\partial T]_\xi$
Hence, $[\partial C_p/\partial p]_{T;\xi} = -T[\partial^2 V/\partial T^2]_{p;\xi}$
or $C_p(T; p_2; \xi) - C_p(T; p_1; \xi) = -\int_{p_1}^{p_2} \{T[\partial^2 V/\partial T^2]_\xi\}\,\mathrm{d}p$

Table 1.5.6. Dependence of entropy on pressure

From Table 1.5.1, $(\partial S/\partial p)_{T;\xi} = -(\partial V/\partial T)_{p;\xi}$
Then, $S(T; p_2; \xi) - S(T; p_1; \xi) = -\int_{p_1}^{p_2} (\partial V/\partial T)_{p;\xi}\,\mathrm{d}p$
or $S(T; p_2; \xi) - S(T; p_1; \xi) = -\int_{p_1}^{p_2} (\alpha(\xi)V)\,\mathrm{d}p$

Table 1.5.7. Dependence of entropy on temperature

From the Gibbs–Helmholtz equation, $(\partial H/\partial T)_{p;\xi} = -T(\partial^2 G/\partial T^2)_{p;\xi}$
But $S = -(\partial G/\partial T)_{p;\xi}$ then, $C_p(\xi) = T(\partial S/\partial T)_{p;\xi}$
Hence, $(\partial S/\partial T)_{p;\xi} = C_p(\xi)/T$
Therefore, $S(T_2; p; \xi) - S(T_1; p; \xi) = \int_{T_1}^{T_2} C_p(\xi)\,\mathrm{d}(\ln T)$

differential with respect to pressure at constant T defines the isothermal compressibility, $\kappa(\xi)$. $\beta(\xi)$ is the isochoric thermal pressure coefficient at constant composition. For stable phases, isothermal compressibilities are positive.

A similar analysis in Table 1.5.5 refers to the dependence of enthalpies and isobaric heat capacities on pressure. The link between the partial derivative of H with respect to pressure and the volumetric properties of a system is particularly interesting.

The entropy of a system is defined by the independent variables T, p and ξ; $S = S[T; p; \xi]$. Here we explore the dependence of S on both p (Table 1.5.6) and T (Table 1.5.7) leading to equations relating the change in entropy to changes in T, p and ξ; Table 1.5.8. Indeed the equation in Table 1.5.6 captures an important flavour

Table 1.5.8. Entropy S as a function of T, p and ξ; $S = S[T;p;\xi]$

$$dS = (\partial S/\partial T)_{p;\xi}\, dT + (\partial S/\partial p)_{T;\xi}\, dp + (\partial S/\partial \xi)_{T;p}\, d\xi$$

But (Tables 1.5.7 and 1.5.1) $(\partial S/\partial T)_{p;\xi} = C_p(\xi)/T$;

$$(\partial S/\partial p)_{T;\xi} = -(\partial V/\partial T)_{p;\xi}$$

Hence, $dS = [C_p(\xi)/T]\, dT - (\partial V/\partial T)_{p;\xi}\, dp + (\partial S/\partial \xi)_{T;p}\, d\xi$

of thermodynamics. The task of determining the dependence of entropy on pressure seems awesome until one spots that this information is readily obtained from volumetric properties.

References to section 1.5

[1] H. Margenau; G. M. Murphy. *Mathematics of Physics and Chemistry*, D. Van Nostrand, Princeton, N.J., 1943, 2nd. edition.
[2] E. F. Caldin. *Chemical Thermodynamics*, Oxford, 1958.

1.6 PROPERTIES OF SYSTEMS AT CONSTANT COMPOSITION AND CONSTANT AFFINITY

We turn to the properties of an aqueous solution containing benzoic acid (represented here by the symbol HA). We have established that at equilibrium the system is at a minimum in Gibbs energy G and the affinity for spontaneous reaction A is zero. The composition can be described in terms of the chemical equilibrium in equation (1.6.1).

$$HA(aq) \rightleftharpoons H^+(aq) + A^-(aq) \qquad (1.6.1)$$

The molalities of solutes at equilibrium are $m^{eq}(HA)$, $m^{eq}(H^+)$ and $m^{eq}(A^-)$ at defined temperature and pressure. In general terms we characterize the composition of the solution by $\xi_1^{eq}(T_1;p_1)$. Here the superscript 'eq' also indicates that under these conditions $A = 0$. Experiment reveals that the composition at equilibrium but at temperature T_2 and pressure p_2 is different; i.e. $\xi_2^{eq}(T_2;p_2)$. The condition, fixed composition, which we introduced in the previous section implies that when the temperature and pressure are changed the composition remains at ξ_1. By removing the 'eq' superscript we acknowledge that the aqueous solution is not at equilibrium at T_2 and p_2 where the composition is ξ_1. $C_p(\xi)$ is the isobaric heat capacity of this solution under conditions where no change occurs in chemical composition. When heat passes from the surroundings into the solution, the temperature rises but no change occurs in composition; i.e. the chemical composition is frozen. The heat capacity $C_p(A = 0)$ corresponds to conditions in which the composition changes to hold the system at zero affinity; i.e. where the compositions are ξ_1^{eq} and ξ_2^{eq} at temperatures T_1 and T_2 respectively at constant pressure. To find out how $C_p(\xi)$ and $C_p(A = 0)$ differ we need to consider the question of the affinity for reaction A in rather more detail.

Table 1.6.1. Affinity

$$A = A[T; p; \xi]$$

From the definition, $G = H - TS$,

$$(\partial G/\partial \xi)_{T;p} = (\partial H/\partial \xi)_{T;p} - T(\partial S/\partial \xi)_{T;p}$$

But (Table 1.3.6) $A = -(\partial G/\partial \xi)_{T;p}$
Hence, $A = -(\partial H/\partial \xi)_{T;p} + T(\partial S/\partial \xi)_{T;p}$

Table 1.6.2. Affinity A as a function of T

From $\mathrm{d}^2S/\mathrm{d}T\,\mathrm{d}\xi = \mathrm{d}^2S/\mathrm{d}\xi\,\mathrm{d}T$, or, $(\mathrm{d}/\mathrm{d}\xi)(\partial S/\partial T) = (\mathrm{d}/\mathrm{d}T)(\partial S/\partial \xi)$
However (Table 1.5.7), $(\partial S/\partial T)_{p;\xi} = C_\mathrm{p}(\xi)/T$,
and (Table 1.6.1) $(\partial S/\partial \xi)_{T;p} = (A/T) + (1/T)(\partial H/\partial \xi)_{T;p}$
Hence $(1/T)(\partial C_\mathrm{p}(\xi)/\partial \xi)_{T;p} = (\mathrm{d}/\mathrm{d}T)[(A/T) + (1/T)(\partial H/\partial \xi)_{T;p})]_{p;\xi}$
Then (Table 1.6.1) $(1/T)(\mathrm{d}/\mathrm{d}T)(\partial H/\partial \xi)_{T;p} = (\mathrm{d}/\mathrm{d}T)(A/T)_{p;\xi}$
$+ (1/T)(\mathrm{d}/\mathrm{d}T)(\partial H/\partial \xi)_{T;p} - (1/T^2)(\partial H/\partial \xi)_{T;p}$
Hence, $(\mathrm{d}/\mathrm{d}T)(A/T)_{p;\xi} = (1/T^2)(\partial H/\partial \xi)_{T;p}$
Further, $-(A/T^2) + (1/T)(\partial A/\partial T)_{p;\xi} = (1/T^2)(\partial H/\partial \xi)_{T;p}$
At equilibrium where $A = 0$, $(\partial A/\partial T)_{p;A=0} = (1/T)(\partial H/\partial \xi)^{\mathrm{eq}}_{T;p}$

The key role played by the affinity for spontaneous change prompts the derivation of equations which link A with other thermodynamic variables. We limit discussion to spontaneous changes in closed systems at fixed temperature and fixed pressure. Affinity A, a dependent variable, is defined by independent variables T, p and ξ (Table 1.6.1). The temperature derivative of (A/T) is related to $(\partial H/\partial \xi)_{T;p}$ (Table 1.6.2).

All spontaneous processes at fixed temperature and pressure lead to a decrease in the Gibbs function G. Formally we represent this conclusion as shown in Fig. 1.6.1. In the top half of the diagram G is plotted against ξ. Both G and ξ are extensive variables. For a larger system, the curve moves to larger values of both G and ξ but overall retains the same shape. A similar figure is sometimes encountered where ξ is shown changing from ($\xi = 0$) to ($\xi = 1.0$), representing complete reaction; in these diagrams ξ has been normalized [1], usually with reference to one product in the stoichiometric equation for chemical reaction.

With increase in ξ, reactants are consumed and products formed, the total G of the system decreasing. The dependence on ξ of the affinity A [$= -(\partial G/\partial \xi)_{T;p}$] is shown in the central part of the same figure. Spontaneous chemical reaction continues until, at ξ^{eq}, G is a minimum and A is zero, the state of thermodynamic equilibrium. Near the minimum in G, the affinity A changes from positive to negative through zero at ξ^{eq}. The condition for bilateral stability [2] of a chemical equilibrium is $(\partial A/\partial \xi)_{T;p} < 0$. Here 'bilateral' means that ξ can, as a result of perturbation, either increase or decrease (i.e. towards reactants or towards products). The perturbation may arise

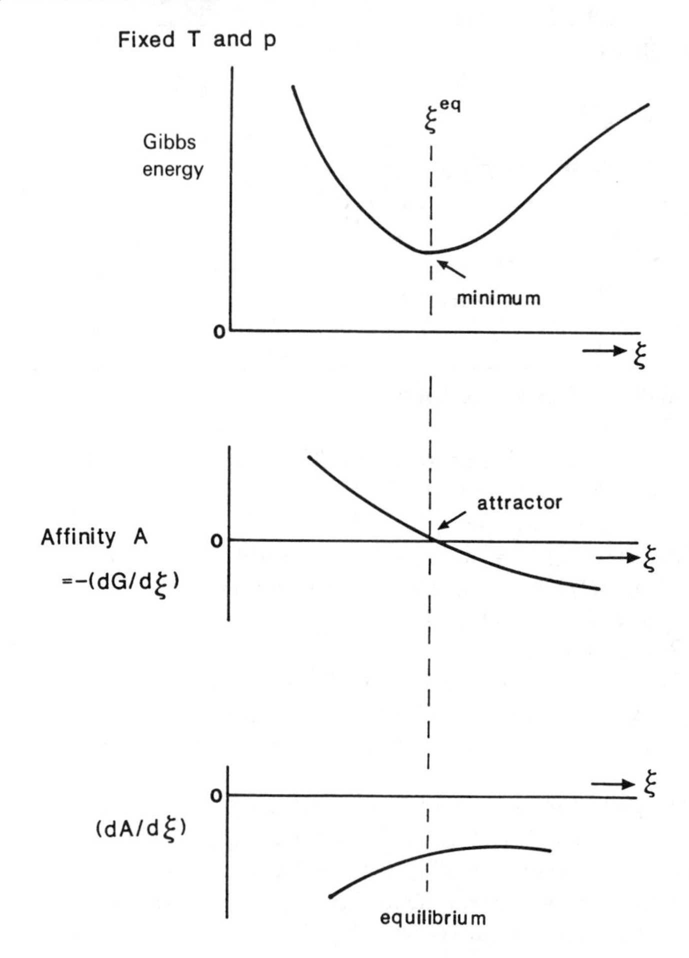

Fig. 1.6.1. Dependence on composition of Gibbs function G, affinity and $(\partial A/\partial\xi)_{T;p}$ about equilibrium for a closed system at fixed T and p.

from a fluctuation in temperature and/or pressure. Alternatively, local fluctuations in composition may occur within microscopic volumes of the system tending to drive the system away from the minimum in G. This is opposed by a non-zero affinity for reaction.

We return to the aqueous solution at temperature T_1 and pressure p_1 in which an equilibrium exists between undissociated and dissociated benzoic acid. The system is at equilibrium (i.e. $A = 0$) and the chemical composition $\xi^{eq} = \xi^{eq}(\text{I})$. The system is displaced by a change of temperature or pressure (thermodynamic reversible process) to a new equilibrium state, $\xi^{eq} = \xi^{eq}(\text{II})$ where again $A = 0$. The change $d\xi$ for this equilibrium displacement is derived in Table 1.6.3 under the constraint that dA is zero: $A(\text{I}) = A(\text{II}) = 0$. At equilibrium (constant T and constant p), G is a minimum. However, there is no accompanying condition associated with the plot of enthalpy H against ξ. Indeed where $(\partial G/\partial\xi)_{T;p} = 0$, $(\partial H/\partial\xi)_{T;p}$ may be zero, positive or negative.

Table 1.6.3. Affinity, A: $(A/T) = (A/T)(T;p;\xi)$

Independent variables, T, p, and ξ; dependent variable (A/T)

$$\mathrm{d}(A/T) = [\partial(A/T)/\partial T]_{p;\xi}\,\mathrm{d}T + [\partial(A/T)/\partial p]_{T;\xi}\,\mathrm{d}p + [\partial(A/T)/\partial\xi]_{T;p}\,\mathrm{d}\xi$$

But $\mathrm{d}(A/T) = (1/T)\,\mathrm{d}A - (A/T^2)\,\mathrm{d}T$.
Hence (Tables 1.5.1 and 1.6.2),

$$(1/T)\,\mathrm{d}A - (A/T^2)\,\mathrm{d}T = (1/T^2)(\partial H/\partial\xi)_{T;p}\,\mathrm{d}T - (1/T)(\partial V/\partial\xi)_{T;p}\,\mathrm{d}p + (1/T)(\partial A/\partial\xi)_{T;p}\,\mathrm{d}\xi$$

Rearrange: $\mathrm{d}A = [\{A + (\partial H/\partial\xi)_{T;p}\}/T]\,\mathrm{d}T - (\partial V/\partial\xi)_{T;p}\,\mathrm{d}p + (\partial A/\partial\xi)_{T;p}\,\mathrm{d}\xi$
Rearrange: $\mathrm{d}\xi = -[\{A + (\partial H/\partial\xi)_{T;p}\}/\{T(\partial A/\partial\xi)_{T;p}\}]\,\mathrm{d}T + [(\partial V/\partial\xi)_{T;p}/(\partial A/\partial\xi)_{T;p}]\,\mathrm{d}p + [1/(\partial A/\partial\xi)_{T;p}]\,\mathrm{d}A$
Hence for an equilibrium change (where $A = 0$),

$$\mathrm{d}\xi(A=0) = -[\{(\partial H/\partial\xi)_{T;p}\}/\{T(\partial A/\partial\xi)_{T;p}\}]\,\mathrm{d}T + [(\partial V/\partial\xi)_{T;p}/(\partial A/\partial\xi)_{T;p}]\,\mathrm{d}p$$

At constant pressure, $(\partial\xi/\partial T)_{p;A=0} = -(\partial H/\partial\xi)_{T;p}/[T(\partial A/\partial\xi)_{T;p}]$
At constant temperature, $(\partial\xi/\partial p)_{T;A=0} = (\partial V/\partial\xi)_{T;p}/(\partial A/\partial\xi)_{T;p}$
where $(\partial A/\partial\xi)_{T;p} < 0$

Table 1.6.4. Van't Hoff's theorem

(a) $(\partial H/\partial\xi) > 0$; endothermic process. Therefore, $(\partial\xi/\partial T)_{p;A=0} > 0$
Conclusion: an increase in T produces an increase in ξ^{eq}
(b) $(\partial H/\partial\xi) < 0$; exothermic process. Therefore, $(\partial\xi/\partial T)_{p;A=0} < 0$
Conclusion: an increase in T produces a decrease in ξ^{eq}.

Table 1.6.5. Le Chatelier's principle

(a) $(\partial V/\partial\xi) > 0$; volume increases with increase in ξ^{eq}.
Hence, $(\partial\xi/\partial p)_{T;A=0} < 0$.
Conclusion: ξ^{eq} decreases with increase in pressure.
(b) $(\partial V/\partial\xi) < 0$; volume decreases with increase in ξ^{eq}.
Hence, $(\partial\xi/\partial p)_{T;A=0} > 0$.
Conclusion: ξ^{eq} increases with increase in pressure.

If $(\partial H/\partial\xi)_{T;p}$ is positive, chemical reaction is endothermic whereas if $(\partial H/\partial\xi)_{T;p}$ is negative, the reaction is exothermic.

In analogous fashion, there is no condition carrying over from the minimum in G to the dependence of V on ξ, i.e. $(\partial V/\partial\xi)_{T;p}$ can be zero, positive or negative at ξ^{eq} where G is a minimum. The equations derived in Table 1.6.3 lead to a quantitative understanding of two theorems [2] (Tables 1.6.4 and 1.6.5). Where chemical reaction

Table 1.6.6. Equilibrium properties

Equilibrium isobaric heat capacity, $C_p(A = 0) = (\partial H/\partial T)_{p;\,A=0}$
From equation (1.4.2), $C_p(A = 0) = C_p(\xi) + (\partial H/\partial \xi)_{T;p}(\partial \xi/\partial T)_{p;\,A=0}$
Equilibrium isobaric thermal expansivity,

$$\alpha(A = 0) = (1/V)(\partial V/\partial T)_{p;\,A=0}$$

Equilibrium isothermal compressibility,

$$\kappa(A = 0) = -(1/V)(\partial V/\partial p)_{T;\,A=0}$$

---

is exothermic, the equilibrium position shifts to favour a system richer in reactants when the temperature increases. If chemical reaction produces a volume increase, an increase in pressure produces a shift in ξ^{eq} towards reactants. In Chapter 4 we comment on the dependence on pressure of the composition at G^{eq}.

The isobaric equilibrium heat capacity $C_p(A = 0)$ means that the compositions at temperature T and $T + \Delta T$ differ, although in both states, $(T; p; \xi^{eq})$ and $\{(T + \Delta T); p; (\xi + \Delta\xi)^{eq}\}$, the affinity for spontaneous change is zero. An equilibrium isobaric thermal expansivity measures the corresponding change in volume; i.e. the difference between $V[T_2; p; \xi_2^{eq}]$ and $V[T_1; p; \xi_1^{eq}]$; (Table 1.6.6). We distinguish, therefore, between properties at constant composition (e.g. $C_p(\xi)$) and properties measured where A is zero; e.g. $C_p(A = 0)$.

References to section 1.6

[1] F. R. Cruikshank; A. J. Hyde; D. Pugh. *J. Chem. Educ.*, 1977, **54**, 288.
[2] I. Prigogine; R. Defay. *Chemical Thermodynamics* (trans. D. H. Everett), Longmans-Green, London, 1954.

1.7 THERMODYNAMIC EQUATIONS OF STATE

For a system at constant composition, the change in thermodynamic energy dU is related to the change in volume and entropy (Table 1.7.1). If the system is held at constant temperature, the dependence of U on V is related to the dependence of p on T at constant volume. The partial derivative $(\partial U/\partial V)_T$ is the internal pressure [1–3], π_i. Then equation (a) of Table 1.7.1 can be written in the simple form for a pure substance j,

$$\pi_i^*(\xi; j) = T[\alpha^*(\xi)/\kappa^*(\xi)] - p \qquad (1.7.1)$$

Here $\pi_i^*(\xi; j)$ is a 'frozen' internal pressure for substance j. Alternatively we derive a related equation for systems which are at all stages at thermodynamic equilibrium (Table 1.7.2). Hence $\pi_i^*(A = 0; j)$ is the equilibrium internal pressure. In fact this quantity is called simply the internal pressure.

Table 1.7.1. Thermodynamic equation of state

For a system at constant composition, $\mathrm{d}U = T\mathrm{d}S - p\mathrm{d}V$
At constant temperature, $(\partial U/\partial V)_{T;\xi} = T(\partial S/\partial V)_{T;\xi} - p$
But (Table 1.5.1) $(\partial S/\partial V)_{T;\xi} = (\partial p/\partial T)_{V;\xi}$
Then, $(\partial U/\partial V)_{T;\xi} = T(\partial p/\partial T)_{V;\xi} - p$;

or $$(\partial U/\partial V)_{T;\xi} = T(\alpha(\xi)/\kappa(\xi)) - p \qquad \text{(a)}$$

Similarly, at equilibrium

$$(\partial U/\partial V)_{T;A=0} = T(\alpha(A=0)/\kappa(A=0)) - p \qquad \text{(b)}$$

Units: $(\partial U/\partial V) = [\mathrm{J}]/[\mathrm{m}^3] = [\mathrm{J\,m^{-1}}][\mathrm{m^{-2}}] = [\mathrm{N\,m^{-2}}] = [\mathrm{Pa}]$

Table 1.7.2. Thermodynamic equation of state

For substance j; $\pi_i(j; A = 0) = T(\partial p/\partial T)_{V_j;A=0} - p$

Table 1.7.3. Compression of liquids

At fixed T and composition
π = ambient pressure, $k_c = \{[V(1; T; \pi) - V(1; T; p)]/V(1; T; \pi)\}$

Table 1.7.4. Tait equation

At constant temperature and composition
$\{V(p_1) - V(p_2)\}/V(p_1) = c \ln\{(b + p_2)/(b + p_1)\}$

Equation (b) of Table 1.7.1 is the thermodynamic equation of state. For perfect gases, the internal pressure is zero (by definition). For real systems (i.e. gases, liquids and solids) internal pressures are non-zero and offer an insight into intermolecular forces within a given system. Internal pressures for a given liquid depend on temperature and pressure, and for liquid mixtures on composition.

At fixed temperature and composition, the molar volume of a liquid depends on pressure, this dependence being characterized by the isothermal compressibility. A liquid is also characterized by the compression, k_c, which measures the relative change in volume for a given increase in pressure (Table 1.7.3). Compressions depend on temperature and pressure.

Numerous equations of state express the dependence of molar volume on pressure. The Tait equation is used for many liquids. Various Tait equations [4] are used but we use (see Chapter 4) the one set out in Table 1.7.4. Here c and b depend on the liquid. Usually pressure p_1 is ambient pressure, $101\,325\,\mathrm{N\,m^{-2}}$.

References to section 1.7

[1] J. H. Hildebrand; R. L. Scott. *The Solubility of Non-electrolytes*, Dover, New York, 1964.

[2] M. J. Blandamer; J. Burgess; A. W. Hakin. *J. Chem. Soc. Faraday Trans. 1*, 1987, **83**, 1783.

[3] M. R. J. Dack. *Chem. Soc. Revs.*, 1975, **4**, 211; *J. Chem. Educ.*, 1974, **51**, 231.

[4] G. S. Kell. *Water—A Comprehensive Treatise*, (ed. F. Franks), Plenum Press, New York, 1972, volume 1, chapter 10.

2

Composition of solutions; chemical potentials

2.1 INTRODUCTION

The main concern of this chapter is a partial molar property called the chemical potential. First, the affinity A is related to the chemical potentials of all substances taking part in a chemical reaction. Second, the chemical potential of each substance is related to the composition of a solution.

The procedure for introducing the chemical potential involves switching our attention from closed to open systems. By way of introduction (Table 2.1.1) we consider an aqueous solution prepared by adding n_2 moles of urea to n_1 moles of water. The thermodynamic energy is defined by four independent variables; S, V, n_1 and n_2. Without examining the practicalities of the process too closely, the chemical potential of urea μ_2 describes the change in thermodynamic energy of the solution when δn_2 moles of urea are added at constant entropy, volume and n_1. In analogous fashion, the chemical potential of water μ_1 measures the change in U when δn_1 moles of water are added at constant S, V and n_2. The important point to note is that these partial molar properties μ_1 and μ_2 describe properties of the two chemical substances in solution in terms of differential changes in a macroscopic thermodynamic property. In these terms a bridge is formed between bulk properties and the properties of

Table 2.1.1. Aqueous solution—chemical potential

$$U = U[S; V; n_1; n_2]$$

Composition: n_1 moles of substance 1, water.
n_2 moles of substance 2, urea

$$dU = (\partial U/\partial S)_{V;n_1;n_2}\, dS + (\partial U/\partial V)_{S;n_1;n_2}\, dV + (\partial U/\partial n_1)_{S;V;n_2}\, dn_1 + (\partial U/\partial n_2)_{S;V;n_1}\, dn_2$$

By definition; $\mu_1 = (\partial U/\partial n_1)_{S;V;n_2}$ and, $\mu_2 = (\partial U/\partial n_2)_{S;V;n_1}$

Also (Table 1.3.4) $T = (\partial U/\partial S)_{V;n_1;n_2}$

and $p = -(\partial U/\partial V)_{S;n_1;n_2}$

Hence, $dU = T\, dS - p\, dV + \mu_1\, dn_1 + \mu_2\, dn_2$

Table 2.1.2. Thermodynamic energy and chemical potential

$$U = U[S; V; n_i]$$

$\mathrm{d}U = (\partial U/\partial S)_{V;n_i}\,\mathrm{d}S + (\partial U/\partial V)_{S;n_i}\,\mathrm{d}V + \Sigma(j = 1; j = \mathrm{i})(\partial U/\partial n_j)_{S;V;n_{i \neq j}}\,\mathrm{d}n_j$
where (Table 1.3.4) $T = (\partial U/\partial S)_{V;n_i}$; $p = -(\partial U/\partial V)_{S;n_i}$
and, by definition*, $\mu_j = (\partial U/\partial n_j)_{S;V;n_{i \neq j}}$
Hence**, $\mathrm{d}U = T\,\mathrm{d}S - p\,\mathrm{d}V + \Sigma(j = 1; j = \mathrm{i})\mu_j\,\mathrm{d}n_j$

*Units: $\mu_j = [\mathrm{J}]/[\mathrm{mol}] = [\mathrm{J\,mol^{-1}}]$
**The fundamental equation of state

Table 2.1.3. Affinity and chemical potential

(Tables 1.3.3 and 2.1.2) Comparison of equations for $\mathrm{d}U$

$$A = -\Sigma(j = 1; j = \mathrm{i})\nu_j\mu_j$$

At equilibrium, $A^{\mathrm{eq}} = -\Sigma(j = 1; j = \mathrm{i})\nu_j\mu_j^{\mathrm{eq}} = 0$

substances making up a system. In other words we reveal to the thermodynamics that a given system is made up of molecules.

The argument is generalized [1] for a system which contains n_1 moles of substance 1, n_2 moles of substance 2, . . . , n_i moles of substance i. We add quite arbitrarily δn_j moles of substance j to the system ($j = 1, 2, 3, \ldots, i$). We assert that, just as for closed systems, there exist two extensive functions of state, the thermodynamic energy U and the entropy S.

The thermodynamic energy U is defined by the independent variables, S, V and n_i (Table 2.1.2). The general differential includes the partial derivative $(\partial U/\partial n_j)$ for substance j at constant S and V where the amounts of all substances other than substance j are constant. For the partial derivatives, $(\partial U/\partial S)_{V;n_i}$ and $(\partial U/\partial V)_{S;n_i}$, the condition that n_i is constant is equivalent to constant ξ (Chapter 1). Comparison of two equations for $\mathrm{d}U$ (Tables 2.1.1 and 1.3.3) yields an equation relating the affinity A and the chemical potentials of substances involved in chemical reaction (Table 2.1.2). The condition $A = 0$ links (Table 2.1.3) chemical potentials at equilibrium.

Reference for section 2.1

[1] I. Prigogine; R. Defay. *Chemical Thermodynamics* (trans. D. H. Everett), Longmans-Green, London, 1954.

2.2 CHEMICAL POTENTIAL AND RELATED PARTIAL MOLAR PROPERTIES

The equation (Table 2.1.2) for the change in thermodynamic energy of an open system for arbitrary change in n_j is combined with the equation defining G in terms of U

Table 2.2.1. Gibbs energy and chemical potentials

From Table 1.3.5, $dG = dU + p\,dV + V\,dp - T\,dS - S\,dT$
Then from Table 2.1.2 (substituting for dU)
$dG = -S\,dT + V\,dp + \Sigma(j = 1; j = i)(\partial U/\partial n_j)_{S;V;n_{i\neq j}}$

Table 2.2.2. Chemical potentials

Dependent variable, $G = G[T; p; n_i]$
$dG = (\partial G/\partial T)_{p;n_i}\,dT + (\partial G/\partial p)_{T;n_i}\,dp + \Sigma(j = 1; j = i)(\partial G/\partial n_j)_{T;p;n_{i\neq j}}\,dn_j$
Comparison of equations for dG;
$S = -(\partial G/\partial T)_{p;n_i}$; $V = (\partial G/\partial p)_{T;n_i}$
and $\mu_j = (\partial U/\partial n_j)_{S;V;n_{i\neq j}} = (\partial G/\partial n_j)_{T;p;n_{i\neq j}}$

Table 2.2.3. Gibbs' theorem—chemical potentials

$\mu_j = (\partial U/\partial n_j)_{S;V;n_{i\neq j}} = (\partial G/\partial n_j)_{T;p;n_{i\neq j}}$
$= (\partial H/\partial n_j)_{S;p;n_{i\neq j}} = (\partial F/\partial n_j)_{T;V;n_{i\neq j}}$
$G = \Sigma(j = 1; j = i)n_j\mu_j$; $U = TS - pV + \Sigma(j = 1; j = i)n_j\mu_j$
$H = TS + \Sigma(j = 1; j = i)n_j\mu_j$; $F = -pV + \Sigma(j = 1; j = i)n_j\mu_j$

(Table 1.3.5). The result (Table 2.2.1) is an equation for the change in G as a function of changes in temperature and pressure together with partial differentials of U defining the chemical potential of substance j.

The condition of constant n_i is equivalent to constant ξ. The partial derivative of G with respect to n_j at constant T, p and $n_{i\neq j}$ equals the chemical potential of substance j (Table 2.2.2).

For a solution of urea in water at fixed temperature and pressure, the chemical potential of urea measures the change in G, δG when δn_j moles of urea are added. Similarly μ_1 measures the change in G, δG when δn_1 moles of water are added. At the same time other thermodynamic properties also change; e.g. volume and entropy.

A similar analysis based on the thermodynamic properties F and H leads to a theorem by Gibbs: the chemical potentials defined by analogous derivatives of H, F and G are equal (Table 2.2.3). Integrated forms of equations for G, F, H and U are summarized (Table 2.2.3) in terms of the amount and chemical potential of each substance j.

Chemical potentials are the senior members of families of partial molar quantities (Table 2.2.4). The partial molar volume of substance j, V_j is the differential change in volume when δn_j moles of substance j are added at constant T and p. V_j is not the molar volume of substance j in the mixture in the sense of the volume occupied by a mole of substance j.

Table 2.2.4. Partial molar properties

Volume $V = \Sigma(j = 1; j = \mathrm{i}) n_j V_j$
Partial molar volume, $V_j = (\partial V / \partial n_j)_{T;p;n_{i \neq j}}$
Isobaric heat capacity $C_p = \Sigma(j = 1; j = \mathrm{i}) n_j C_{pj}$
Partial molar isobaric heat capacity, $C_{pj} = (\partial C_p / \partial n_j)_{T;p;n_{i \neq j}}$
Enthalpy $H = \Sigma(j = 1; j = \mathrm{i}) n_j H_j$
Partial molar enthalpy, $H_j = (\partial H / \partial n_j)_{T;p;n_{i \neq j}}$
Gibbs energy $G = \Sigma(j = 1; j = \mathrm{i}) n_j \mu_j$
Chemical potential, $\mu_j = (\partial G / \partial n_j)_{T;p;n_{i \neq j}}$
Entropy $S = \Sigma(j = 1; j = \mathrm{i}) n_j S_j$
Partial molar entropy, $S_j = (\partial S / \partial n_j)_{T;p;n_{i \neq j}}$

Units: $C_{pj} = [\mathrm{J\,mol^{-1}\,K^{-1}}]$; $H_j = [\mathrm{J\,mol^{-1}}]$
$\mu_j = [\mathrm{J\,mol^{-1}}]$; $S_j = [\mathrm{J\,K^{-1}\,mol^{-1}}]$

The change in volume of an empty egg carton when an egg is added is zero. Therefore the partial molar volume of eggs in an egg carton is zero. For a carton designed to hold six eggs, this state of affairs continues until the carton is full. Beyond that mole fraction of eggs, x_{egg}, the dependence of V_{egg} on x_{egg} is complicated.

2.3 DESCRIPTIONS OF SYSTEMS

In the previous section we commented on the properties of a solution prepared using n_1 moles of water and n_2 moles of urea at ambient pressure and 298 K. Let us imagine for the moment that we have just such an aqueous solution ($n_1 \gg n_2$) contained in a flask which stands on the bench in front of us. In reviewing the chemistry of this system, we could attempt to understand the interactions between water and urea molecules. We could cite treatments of liquid water which suggest that in aqueous solutions there are water molecules which are hydrogen bonded to differing extents. We could also consider the possibility that a fraction of the urea molecules exist as dimers. While we debate these issues and formulate increasingly complex and wildly different models for the system, the solution sits on the bench, probably perplexed (in anthropomorphic terms) by the intensity of the debate going on around it. Clearly the volume of a solution does not change as we alter our model for the interactions between molecules. The question emerges as to which properties are model-dependent. In other words we want to identify those properties we should measure in order to test a given model. To develop these points we turn our attention to a closed system at equilibrium and fixed temperature and pressure containing methyl alcohol and tetrachloromethane.

In broad terms, the Gibbs energy is related to the composition and chemical potentials of each substance in a system. Further, G is a function of state and, therefore, is independent of how we, as observers, choose to describe the system. We consider two descriptions for the system described in Table 2.3.1.

Table 2.3.1. Description of a system

Closed system; fixed T and p.
System prepared using n_1 moles of CCl_4 and n_2 moles of CH_3OH.
Condition: equilibrium.
Description I Solution of monomeric CH_3OH molecules in CCl_4

$$G^{eq} = n_1 \mu^{eq}(CCl_4; \text{system}) + n_2 \mu^{eq}(CH_3OH; \text{system})$$

where substance 1 = CCl_4 and substance 2 = CH_3OH

Description II $\underset{n_2^{eq}}{2CH_3OH} \rightleftharpoons \underset{n_d^{eq}}{(CH_3OH)_2}$

Solution of n_2^{eq} moles of alcohol monomer
and n_d^{eq} moles of alcohol dimer.
Hence, $G^{eq} = n_1 \mu^{eq}(CCl_4; \text{system}) + n_2^{eq} \mu^{eq}(CH_3OH; \text{system}) + n_d^{eq} \mu^{eq}((CH_3OH)_2; \text{system})$
At equilibrium (Table 2.1.3), $A^{eq} = 0 = \Sigma(j = 1; j = i)\{v_j \mu_j\}^{eq}$
where $v((CH_3OH)_2) = +1$ and $v(CH_3OH) = -2$

$$2\mu^{eq}(CH_3OH; \text{system}) = \mu^{eq}((CH_3OH)_2; \text{system})$$

Then $G^{eq} = n_1 \mu^{eq}(CCl_4; \text{system}) + [n_2^{eq} + 2n_d^{eq}]\mu^{eq}(CH_3OH; \text{system})$
From the conservation of mass, $n_2^{eq} + 2n_d^{eq} = n_2$
Then, $G^{eq} = n_1 \mu^{eq}(CCl_4; \text{system}) + n_2 \mu^{eq}(CH_3OH; \text{system})$
(cf. description I)

In description I, methyl alcohol is present as monomeric alcohol molecules. In description II, the alcohol is present as monomers and dimers in chemical equilibrium. Both descriptions lead to identical G^{eq}. Indeed, no matter how complicated the description of a system we offer, the resultant expression for G^{eq} must be consistent with other simpler descriptions. In general terms description and definition play an important role in understanding the chemistry of a given system.

2.4 GIBBS FUNCTIONS AND DESCRIPTIONS

The question of description is vital to our understanding of the interplay between experiment and definition of thermodynamic variables. To give point to the discussion we consider a closed one-component homogeneous system which contains n_x moles of substance X at temperature T and pressure p. Experiment confirms that the system is in equilibrium with the surroundings. Moreover this experiment prompts the conclusion that no chemical reaction has taken place and that the system comprises simply n_x moles of X (description I, Table 2.4.1). The molar volume, $V_x^* = (V/n_x)$.

Another experiment prompts the conclusion that the system at temperature T and pressure p comprises two substances X and Y in chemical equilibrium. In description II, therefore, the composition is given by $n_x^{eq}(\text{II})$ and $n_y^{eq}(\text{II})$ (Table 2.4.2). The question

Table 2.4.1. System A—description I:

Composition: n_x moles of substance X
Conditions: equilibrium at temperature T and pressure p

$$G^{eq}(I) = G[T; p; n_x] \qquad \mu_x^{eq}(I) = (\partial G/\partial n_x)_{T;p}^{eq}$$

Table 2.4.2. System A—description II

System prepared as n_x moles of X
Conditions: equilibrium at temperature T and pressure p

$$X \Leftarrow\!\!\Rightarrow Y$$

$$n_x^{eq}(II) \qquad n_y^{eq}(II)$$

Then, $G^{eq}(II) = G[T; p; n_x^{eq}; n_y^{eq}]$ and $\mu_x^{eq} = (\partial G(II)/\partial n_x)_{T;p}^{eq}$

Table 2.4.3. Partial derivatives of G—chemical potentials

System: Tables 2.4.1 and 2.4.2. Conditions: fixed T and p.
$(\partial G/\partial n_x)_A = (\partial G/\partial n_x)_\xi - (\partial A/\partial n_x)_\xi(\partial \xi/A)_{n_x}(\partial G/\partial \xi)_{n_x}$
But $A = -(\partial G/\partial \xi)_{T;p}$
$(\partial G/\partial n_x)_A = (\partial G/\partial n_x)_\xi + (\partial A/\partial n_x)_\xi(\partial \xi/\partial A)_{n_x} A$
At equilibrium, $A = 0$. Then $(\partial G/\partial n_x)_{A=0} = (\partial G/\partial n_x)_{\xi eq}$

arises as to the impact of descriptions I and II, on how we characterize the system. The answer emerges from a consideration of a given extensive property P and the impact of adding δn_x moles of X. In terms of description II, we envisage two limiting cases. In the first, the chemical equilibrium between X and Y does not respond, yielding an 'instantaneous' or 'frozen' property characterized by $(\partial P/\partial n_x)$ at the original ξ^{eq} but now we should denote the extent of reaction in the system simply by ξ because $A \neq 0$. In the second case the chemical reaction responds to hold the system at chemical equilibrium, yielding a property characterized by $(\partial P/\partial n_x)$ at a new equilibrium extent of reaction where $A = 0$. Isobaric heat capacities $C_p(\xi^{eq})$ and $C_p(A = 0)$ characterize the two limiting cases. By way of contrast we showed in Table 2.3.1 that the property G^{eq} is not a function of how we describe the system, i.e. $G^{eq}(I) = G^{eq}(II)$ for system A at temperature T and pressure p.

The partial differential $(\partial G/\partial n_x)_\xi$ characterizes the differential dependence of G on n_x for a system containing n_x moles of substance X at temperature T and pressure p and at fixed ξ. The latter derivative is related to the partial derivative $(\partial G/\partial n_x)$ at fixed T and p and at constant affinity A (Table 2.4.3). In this table we show that partial differentials of G with respect to n_x at $A = 0$ and at ξ^{eq} are equal. In other words, the

Table 2.4.4. Partial derivatives of G—volume

Systems: Tables 2.4.1 and 2.4.2. At temperature T,

$$(\partial G/\partial p)_A = (\partial G/\partial p)_\xi - (\partial A/\partial p)_\xi(\partial \xi/\partial A)_p(\partial G/\partial \xi)_p$$

Hence, $(\partial G/\partial p)_{A=0} = (\partial G/\partial p)_{\xi^{eq}} = V$

Table 2.4.5. Partial derivatives of G—enthalpy and entropy

Systems: Tables 2.4.1 and 2.4.2. At pressure p,

$$(\partial G/\partial T)_A = (\partial G/\partial T)_\xi - (\partial A/\partial T)_\xi(\partial \xi/\partial A)_T(\partial G/\partial \xi)_T$$

But $A = -(\partial G/\partial \xi)_{T;p}$. Then, $(\partial G/\partial T)_{A=0} = (\partial G/\partial T)_{\xi^{eq}} = -S$

$$(\partial(G/T)/\partial T)_A = (\partial(G/T)/\partial T)_\xi - (\partial A/\partial T)_\xi(\partial \xi/\partial A)_p(\partial(G/T)/\partial \xi)_p$$

$$(\partial(G/T)/\partial \xi)_T = (1/T)(\partial G/\partial \xi)_T = -A$$

Then, $(\partial(G/T)/\partial T)_{A=0} = (\partial(G/T)/\partial T)_{\xi^{eq}} = -H/T^2$

chemical potentials of X in system A are the same for descriptions I and II at equilibrium. This identity emerges because at equilibrium $(\partial G/\partial \xi)_{T;p}$ is zero. In other words G is a minimum at equilibrium and using description II, $\mu_x^{eq} = \mu_y^{eq}$. In Chapter 3, we comment on the description of the composition of solutions using equilibrium constants. In other words, these constants characterize systems at minima in G (at fixed T and p).

The condition $(\partial G/\partial \xi)^{eq} = 0$ at fixed T and p, also carries over to the partial derivatives with respect to pressure at ξ^{eq} and at $A = 0$ (Table 2.4.4). We would be surprised if it were otherwise because the volume of system A (at equilibrium and fixed T and p) is not dependent on how we describe the composition. Similar conclusions follow for the derivative of G with respect to temperature (at fixed pressure) at ξ^{eq} and at $A = 0$ (Table 2.4.5). The entropy S and the enthalpy H are not functions of the description. In effect μ_x^{eq}, V^{eq}, S^{eq} and H^{eq} are given by first derivatives of the Gibbs function. However, the equilibrium volume V^{eq} is not at a minimum in a plot showing the dependence of volume V on composition. Similar comments apply to H^{eq}, S^{eq}, and C_p^{eq}. Therefore we anticipate problems when we attempt to understand the dependences of G^{eq} on temperature and on pressure (Chapters 4–6).

The second derivative of G, once with respect to pressure and once with respect to n_x, leads to the partial molar volume of substance X (Table 2.4.6). In the latter case, the triple product term includes the partial derivative $(\partial V/\partial \xi)$ at fixed T, p, and n_x. At equilibrium (at fixed T and p), this quantity is not zero. Returning therefore to description II of system A (Table 2.4.2), we start with a system at equilibrium where the composition is characterized by ξ^{eq} and the volume is V^{eq}. We add δn_x moles of X but the number of moles of Y is not allowed to change; the system remains frozen,

Table 2.4.6. Partial derivatives of V

System prepared using n_x moles of substance X.
Conditions: fixed temperature T and pressure p.
Partial molar volume

$$(\partial V/\partial n_x)_A = (\partial V/\partial n_x)_\xi - (\partial A/\partial n_x)_\xi (\partial \xi/\partial A)_{n_x} (\partial V/\partial \xi)_{n_x}$$

Hence, $V_x(A = 0) = V_x(\xi^{eq}) - (\partial A/\partial n_x)_{\xi^{eq}} (\partial \xi/\partial A)^{eq}_{n_x} (\partial V/\partial \xi)^{eq}_{n_x}$
Temperature dependence (at fixed n_x and p)

$$(\partial V/\partial T)_A = (\partial V/\partial T)_\xi - (\partial A/\partial T)_\xi (\partial \xi/\partial A)_T (\partial V/\partial \xi)_T$$

[divide by V] $\alpha(A = 0) = \alpha(\xi^{eq}) - (1/V)(\partial A/\partial T)_{\xi^{eq}} (\partial \xi/\partial A)^{eq}_T (\partial V/\partial \xi)^{eq}_T$
Pressure dependence (at fixed T and n_x)

$$\kappa(A = 0) = \kappa(\xi^{eq}) + (1/V)(\partial A/\partial p)_{\xi^{eq}} (\partial \xi/\partial A)^{eq}_p (\partial V/\partial \xi)^{eq}_p$$

Further from equation (b) of Table 1.5.1,

$$\kappa(A = 0) = \kappa(\xi^{eq}) - (1/V)(\partial \xi/\partial A)^{eq}_p [(\partial V/\partial \xi)^{eq}_T]^2$$

at ξ^{eq}. The volume change from V^{eq} to $V^{eq} + \delta V$ yields the partial molar volume of X at ξ^{eq}. In a second experiment when δn_x moles of substance X are added, the chemical reaction responds to hold the affinity A at zero. The volume changes from V^{eq} to $V^{eq} + \delta V$ at $A = 0$ to yield the equilibrium partial molar volume. The two partial molar volumes $V_x(\xi^{eq})$ and $V_x(A = 0)$ are related (Table 2.4.6). In practice we measure (indirectly via the densities) $V_x(A = 0)$ which we label V_x. In terms of description I (Table 2.4.1), this is the molar volume of X, V_x^*. If we adopt description II, $V_x(A = 0)$ can be understood in terms of (i) a frozen contribution and (ii) a shift in composition, the first and second terms respectively on the right-hand side of the equation for $V_x(A = 0)$.

Similar arguments emerge for the temperature and pressure derivatives of V which lead in turn to (i) frozen and equilibrium expansibilities and (ii) frozen and equilibrium isothermal compressibilities. Interpretation of the two compressibilities $\kappa(\xi^{eq})$ and $\kappa(A = 0)$ depends on the chemistry involved.

In analogous fashion we identify frozen $\kappa_S(\xi^{eq})$ and equilibrium $\kappa_S(A = 0)$ isentropic compressibilities which are important in the application of ultrasonic relaxation techniques [1] in the study of fast reactions. At high frequencies, a system is characterized by $\kappa_S(\xi^{eq})$ because the pressure changes too fast for the chemical reaction to follow. At low frequencies, the changes in composition approximate to the condition $A = 0$, leading to the characterization of a system by $\kappa_S(A = 0)$. With reference to $\kappa(\xi^{eq})$ and $\kappa(A = 0)$, if the pressure changes, $p \rightarrow p + \delta p$ so rapidly that the composition remains at ξ^{eq}, the volume change is smaller than in the case of a gradual change in pressure.

Table 2.4.7. Partial derivatives of enthalpy H

System: see Table 2.4.6
Partial molar enthalpy (at fixed p and T)

$$(\partial H/\partial n_x)_A = (\partial H/\partial n_x)_\xi - (\partial A/\partial n_x)_\xi (\partial \xi/\partial A)_{n_x} (\partial H/\partial \xi)_{n_x}$$

Hence,

$$H_x(A=0) = H_x(\xi^{eq}) - (\partial A/\partial n_x)_{\xi eq} (\partial \xi/\partial A)^{eq}_{n_x} (\partial H/\partial \xi)^{eq}_{n_x}$$

Pressure dependence (at fixed T and n_x)

$$(\partial H/\partial p)_{A=0} = (\partial H/\partial p)_{\xi eq} - (\partial A/\partial p)^{eq}_\xi (\partial \xi/\partial A)^{eq}_p (\partial H/\partial \xi)^{eq}_p$$

Table 2.4.8. Isobaric heat capacities

At fixed p and n_x

$$C_p(A=0) = C_p(\xi^{eq}) - (\partial A/\partial T)_{\xi eq} (\partial \xi/\partial A)^{eq}_{n_x} (\partial H/\partial \xi)^{eq}_p$$

From Table 1.6.2:

$$C_p(A=0) = C_p(\xi^{eq}) - (1/T)(\partial \xi/\partial A)^{eq}_{n_x} [(\partial H/\partial \xi)^{eq}_p]^2$$

Derivatives of enthalpies H with respect to n_x identify fixed and equilibrium partial molar enthalpies (Table 2.4.7). These distinctions arise because at equilibrium $(\partial H/\partial \xi)$ is not zero. Therefore the partial molar enthalpy of X in a system at equilibrium can be interpreted in two ways.

Using description I (Table 2.4.1), this quantity is the molar enthalpy of X, H_x^*. Using description II (Table 2.4.2), we might understand $H_x(A=0)$ in terms of $H_x(\xi^{eq})$ and a triple product term involving $(\partial H/\partial \xi)_{T;p}$, the enthalpy of reaction. The total enthalpy of the system H is description-independent. But by measuring the dependence on temperature of the equilibrium composition (at G^{eq}) we can probe the enthalpy of reaction (Chapters 5 and 6).

The temperature derivatives of H lead to two isobaric heat capacities, $C_p(\xi^{eq})$ and $C_p(A=0)$ (Table 2.4.8). The analysis is more complicated when we turn attention to the partial molar heat capacities of, for example, X in system A using description II (Table 2.4.2). The third derivative of G is involved, twice with respect to temperature and once with respect to n_x. The complexity is clear if we start with the equation for partial molar enthalpies given in Table 2.4.7.

Differentiation with respect to T identifies two quantities $C_{px}(A=0)$ and $C_{px}(\xi^{eq})$ together with an alarming differential of a triple product term. No condition, based on the Second Law, determines the signs and magnitudes of $C_{px}(\xi^{eq})$ and $C_{px}(A=0)$

(Table 2.4.8). One consequence is that the isobaric heat capacities of reaction and activation are difficult to interpret.

This process of repeated differentiation of the Gibbs energy is conveniently tracked [2] using Fig. 1.3.1, which uses G as the reference property. The concentric rings I, II, and III mark the number of differential operations. The arguments presented above indicate that the task of interpreting derived thermodynamic properties becomes more difficult as we move away from the centre because no analogous condition is attached to other functions of state comparable to the requirement that at equilibrium $(\partial G/\partial \xi)_{T;p}$ is zero.

References to section 2.4

[1] M. J. Blandamer. *Introduction to Chemical Ultrasonics*, Academic Press, New York, 1973.
[2] M. J. Blandamer; J. Burgess. *Educ. in Chem.*, 1987, **24**, 85.

2.5 GIBBS–DUHEM AND GIBBS–HELMHOLTZ EQUATIONS

The chemical potentials of all substances in a solution are linked; they cannot be varied independently. In a sense there is a communication within a system such that the chemical potential of substance j cannot change independently of the chemical potentials of all other i substances (where $i \neq j$). This communication is quantified by the Gibbs–Duhem equation. Just as in a human family, its members are sensitive to the sadness or happiness of one of their group; often they exert a calming influence —Gibbs–Duhem communication.

Table 2.5.1. Gibbs–Duhem equation—Gibbs function

$$G = \Sigma(j = 1; j = \mathrm{i}) n_j \mu_j$$

Then, $\mathrm{d}G = \Sigma(j = 1; j = \mathrm{i})(n_j\,\mathrm{d}\mu_j + \mu_j\,\mathrm{d}n_j)$
But from Table 2.2.1, $\mathrm{d}G = -S\,\mathrm{d}T + V\,\mathrm{d}p + \Sigma(j = 1; j = \mathrm{i})\mu_j\,\mathrm{d}n_j$
Hence, $-S\,\mathrm{d}T + V\,\mathrm{d}p - \Sigma(j = 1; j = \mathrm{i}) n_j\,\mathrm{d}\mu_j = 0$

Table 2.5.2. Gibbs–Helmholtz equation (equilibrium)

Gibbs function. Equilibrium displacement; $A = 0$; fixed pressure.
From Table 1.3.5, $S = -(\partial G/\partial T)_{A=0}$.
But $G = H - TS$; or, $H = G + TS$
or, $H = G - T(\partial G/\partial T)_{A=0}$.
Consider $\mathrm{d}(G/T)/\mathrm{d}T = -(G/T^2) + (1/T)(\mathrm{d}G/\mathrm{d}T)$
or $\mathrm{d}(G/T)/\mathrm{d}T = -(1/T^2)[G - T(\mathrm{d}G/\mathrm{d}T)]$
Therefore,* $(\partial(G/T)/\partial T)_{A=0} = -H/T^2$

*Units: $[\mathrm{J}][\mathrm{K}]^{-1}/[\mathrm{K}] = [\mathrm{J}]/[\mathrm{K}]^2$

Table 2.6.1. Partial molar properties

From Table 1.4.1 $G = H - TS$
Then $(\partial G/\partial n_j)_{T;p;n_{i \neq j}} = (\partial H/\partial n_j)_{T;p;n_{i \neq j}} - T(\partial S/\partial n_j)_{T;p;n_{i \neq j}}$
$H_j = (\partial H/\partial n_j)_{T;p;n_{i \neq j}}$; and $S_j = (\partial S/\partial n_j)_{T;p;n_{i \neq j}}$
Then, $\mu_j = H_j - TS_j$
At constant temperature, $\mathrm{d}^2G/\mathrm{d}n_j\,\mathrm{d}p = \mathrm{d}^2G/\mathrm{d}p\,\mathrm{d}n_j$
Hence, $V_j = (\partial V/\partial n_j)_{T;p} = (\partial \mu_j/\partial p)_T$
Gibbs–Helmholtz equation: $(\partial(\mu_j/T)/\partial T)_p = -H_j/T^2$

Also, $C_{pj} = (\partial H_j/\partial T)_p$

The Gibbs–Duhem equation is derived in Table 2.5.1 using the thermodynamic potential function G. For systems at fixed temperature and pressure,

$$\Sigma(j = 1; j = \mathrm{i}) n_j\,\mathrm{d}\mu_j = 0 \tag{2.5.1}$$

For an equilibrium transformation, the differential gradient of (G/T) with respect to temperature T at constant pressure yields information concerning the enthalpy H at temperature T (Table 2.5.2).

2.6 EQUATIONS RELATING PARTIAL MOLAR PROPERTIES

Starting with the definition of G in terms of H and TS (Table 1.4.1) we differentiate the equation with respect to n_j at constant T, p and $n_{i \neq j}$ to obtain an equation relating the chemical potential, partial molar enthalpy and partial molar entropy of substance j. Second differentials of G yield equations relating μ_j with S_j, V_j and H_j (Tables 2.6.1 and 2.6.2). We note in particular the equation for V_j. The chemical potentials of substance j at two pressures, p_1 and p_2 are related (at fixed temperature):

$$\mu_j(p_2) - \mu_j(p_1) = \int_{p_1}^{p_2} V_j\,\mathrm{d}p \tag{2.6.1}$$

In general, this integral cannot be evaluated because for real systems the relationship between V_j and p is unknown. The pressure derivative of the partial molar enthalpy is related to the temperature derivative of the partial molar volume (Table 2.6.2).

Gradually we have extended the First and Second Laws from statements about macroscopic properties to equations describing partial molar properties. Further the equilibrium state is the reference state for closed systems where at fixed T and p, G is a minimum and A is zero. The point at which A is zero operates as an attractor to the system in the sense that systems in the neighbourhood of this point are drawn towards it (Fig. 1.6.1). The plot of G against ξ is a plot of the quantity $\{\Sigma(j = 1; j = \mathrm{i}) n_j \mu_j\}$ against ξ. Similarly the plot of A against ξ is a plot of $-\Sigma(j = 1; j = \mathrm{i}) v_j \mu_j$ against ξ. Moreover A is intensive and so independent of the size of the system.

Table 2.6.2. Partial molar volume and partial molar enthalpy

At fixed composition, $\mu_j = \mu_j(T; p)$. Then, $d\mu_j^2/dT\,dp = d\mu_j^2/dp\,dT$
Hence, $-(\partial S_j/\partial p)_T = (\partial V_j/\partial T)_p$. But, $\mu_j = H_j - TS_j$
Then at constant temperature, $(\partial \mu_j/\partial p) = (\partial H_j/\partial p) - T(\partial S_j/\partial p)$
Hence, $V_j = (\partial H_j/\partial p) + T(\partial V_j/\partial T)_p$;
or $(\partial H_j/\partial p) = V_j - T(\partial V_j/\partial T)_p$
Therefore,* at pressures p_1 and p_2, (fixed T)

$$H_j(p_2) - H_j(p_1) = \int_{p_1}^{p_2} [V_j - T(\partial V_j/\partial T)]\,dp$$

*Units: $[\text{J mol}^{-1}] = \{[\text{m}^3\,\text{mol}^{-1}] - [\text{K}][\text{m}^3\,\text{mol}^{-1}]/[\text{K}]\}[\text{N m}^{-2}] = [\text{m}^3\,\text{mol}^{-1}][\text{N m}^{-2}]$

At equilibrium,

$$\Sigma(j = 1; j = i)\nu_j\mu_j^{eq} = 0 \tag{2.6.2}$$

Hence if we can relate μ_j^{eq} to intensive composition variables, the equilibrium state corresponding to the minimum in G can be characterized. Away from equilibrium, all spontaneous changes lead to a decrease in G. The system samples the gradient $(dG/d\xi)$ at a given ξ through local fluctuations in composition within each microscopic volume of the system. Fluctuations which result in an overall decrease in G grow at the expense of other changes. At the minimum in G, local fluctuations result in an increase in G and are therefore opposed, bringing the system back to the minimum. The overview offered by Fig. 1.6.1 represents the limit of strictly thermodynamic information. Thermodynamics predicts neither compositions of systems at equilibrium nor rates of change.

Our attention now turns to the task of relating chemical potentials to composition. Some of the elegance of thermodynamics is lost. Initially we consider a solution comprising a single solute in a solvent. In describing the properties of systems, various reference and standard states are defined. Standard pressure p^0 equals $10^5\,\text{N m}^{-2}$. Ambient pressure is 101 325 N m^{-2} which is close to p^0. The difference between these pressures is ignored except in the most precise studies. There is no agreed standard temperature but 298.15 K is a common reference temperature.

2.7 LIQUID MIXTURES AND SOLVENTS

The chemical potential of a gas j at pressure p and temperature T is defined in terms of a standard chemical potential, $\mu_j^0(g; T)$ (Table 2.7.1). The latter is the chemical potential of pure gas j exhibiting ideal gas behaviour at temperature T and standard pressure p^0. If gas j is an ideal gas, the integral term in molar volume V_j^* vanishes.

A given liquid mixture comprises i substances at temperature T and pressure p. By definition [1] the chemical potential of substance j, one of the i substances, is related to the mole fraction composition of the liquid mixture (Table 2.7.2). $\mu_j(\text{mix}; T; p)$ is the chemical potential of substance j at mole fraction x_j in a mixture where the mole fractions of components are described by the collective term x_i. $\mu_j^*(1; T; p)$ is the chemical potential of pure liquid j at the same temperature and pressure. In these

Table 2.7.1. Chemical potential of an ideal gas, substance j

At temperature T
Number of moles $= n_j$; volume, V; pressure p; $V_j^* = V/n_j$
By definition,

$$\mu_j(g;p) = \mu_j^0(g;p^0) + RT\ln(p/p^0) + \int_0^p (V_j^* - RT/p)\,dp$$

If gas j is ideal, $\mu_j(pfg;p) = \mu_j^0(g;p^0) + RT\ln(p/p^0)$

Table 2.7.2. Liquid mixtures—chemical potentials

Temperature T; pressure p
Liquid mixture with i components including liquid j;
By definition; $\mu_j(\text{mix}) = \mu_j^*(1) + RT\ln(x_j f_j)$
where limit $(x_j \to 1.0) f_j = 1.0$ at all T and p
$\mu_j^*(1)$ is the chemical potential of pure liquid j at same T and p.

$$\text{limit}\,(x_j \to 1)\mu_j(\text{mix}) = \mu_j^*(1)$$

Table 2.7.3. Liquid mixtures—volumes and enthalpies

$$V_j = (\partial\mu_j/\partial p)_T$$

From Table 2.7.2 $V_j(\text{mix}; T; p) = V_j^*(1; T; p) + RT(\partial\ln f_j/\partial p)_T$
From the definition of f_j, limit $(x_j \to 1.0) V_j(\text{mix}; T; p) = V_j^*(1; T; p)$
Also $V_j^*(1; T; p^0) = V_j^0(1; T)$

From Table 2.6.1 $(\partial(\mu_j/T)/\partial T)_p = -H_j/T^2$
Then (Table 2.7.1) $H_j(\text{mix}; T; p) = H_j^*(1; T; p) - RT^2(\partial\ln f_j/\partial T)_p$
From the definition of f_j,

$$\text{limit}\,(x_j \to 1.0) H_j(\text{mix}; T; p) = H_j^*(1; T; p)$$

$$\text{Also } H_j^*(1; T; p^0) = H_j^0(1; T)$$

terms f_j expresses [2] the extent to which the properties of substance j in the mixture deviate from (thermodynamic) ideal. In an ideal mixture, $f_j = 1.0$ for all i substances at all compositions. Using the definition of f_j, we establish equations for the partial molar volume and enthalpy as functions of composition (Table 2.7.3).

From the definition of f_j, V_j extrapolated to $x_j = 1.0$ is the molar volume of pure liquid j at the same T and p. Similarly, the partial molar enthalpy in the limit $(x_j \to 1.0)$ is the molar enthalpy of pure liquid j at the same T and p. These extrapolations

Table 2.7.4. Liquid mixtures—standard chemical potentials

From $(\partial\mu_j^*/\partial p)_T = V_j^*$, at temperature T

Hence, $\mu_j(\text{mix};p) = \mu_j^*(1;p^0) + RT\ln(x_j f_j) + \int_{p^0}^{p} V_j^*\,\mathrm{d}p$

where $\mu_j^*(1;p^0) = \mu_j^0(1)$, the standard chemical potential of pure liquid j at temperature T and standard pressure p^0.

Table 2.7.5. Chemical potential of a solvent in a solution

Solution = i solutes + solvent 1; fixed T; fixed p

At fixed temperature

Method I: $\mu_1(\text{sln};T;p) = \mu_1^0(1;T) + RT\ln(x_1 f_1) + \int_{p^0}^{p} V_1^*(1)\,\mathrm{d}p$

where limit $(x_1 \to 1.0)f_1 = 1.0$, and $\mu_1^0(1;T)$ = chemical potential of pure liquid 1 at temperature T and pressure p^0.

Method II*:

$$\mu_1(\text{sln};T;p) = \mu_1^0(1;T) - \phi RTM_1 m_i + \int_{p^0}^{p} V_1^*(1)\,\mathrm{d}p$$

where ϕ = practical osmotic coefficient for the solvent, substance 1.

$$m_i = \Sigma(j=2;j=\text{i})m_j;\ m_j = n_j/n_1 M_1 = n_j/w_1$$

*Units: $[\text{J mol}^{-1}] = [\text{J mol}^{-1}] - [1][\text{J mol}^{-1}\,\text{K}^{-1}][\text{K}][\text{kg mol}^{-1}][\text{mol kg}^{-1}]$
$+ [\text{m}^3\,\text{mol}^{-1}][\text{N m}^{-2}]$

establish two important criteria [3, 4] for equations relating chemical potentials to composition. We require that meaningful H_j and V_j are obtained in an extrapolated limit.

In the equation (Table 2.7.2) for the chemical potential of substance j in a mixture, the reference is the chemical potential of pure liquid j at the same T and p. The standard state for substance j is pure liquid j at temperature T and standard pressure, p^0 (Table 2.7.4).

A given system contains n_j moles of each chemical substance 1, 2, 3, ..., k where $n_1 \gg n_2, n_3, \ldots, n_k$. Substance 1 is the solvent and all other substances are solutes. The solution comprises $k - 1$ solutes in a single solvent 1. Two equations relate the chemical potential of the solvent in solution to the composition (Table 2.7.5 and 2.7.6). The first equation is carried over from treatments of liquid mixtures (Table 2.7.2). The standard states for solvents are corresponding pure liquids at the same temperature and standard pressure, p^0; deviations in the properties of solvents from ideal are described by the rational activity coefficient, f_1. The second method uses the same standard state for solvents but defines a practical osmotic coefficient, ϕ, which

Table 2.7.6. Solvent in solution

At fixed temperature, $(\partial\mu_1/\partial p) = V_1^*$

Hence, $\mu_1^*(1;p) - \mu_1^0(1) = \int_{p^0}^{p} V_1^*(1)\,\mathrm{d}p$

From Table 2.7.5:

Method I $\mu_1(\text{sln};T;p) = \mu_1^*(1;T;p) + RT\ln(x_1 f_1)$

Method II $\mu_1(\text{sln};T;p) = \mu_1^*(1;T;p) - \phi RTM_1\Sigma(j=2;j=k)m_j$

For ideal systems; $\mu_1(\text{sln};T;p) = \mu_1^*(1;T;p) - RTM_1\Sigma(j=2;j=k)m_j$

is unity for ideal solutions. The chemical potentials of solvents in ideal solutions are lower than those of the corresponding pure liquids.

References to section 2.7

[1] M. L. McGlashan. *Chemical Thermodynamics*, Academic Press, London, 1979.
[2] J. S. Rowlinson. *Liquids and Liquid Mixtures*, London, Butterworths, 3rd edn, 1982.
[3] G. N. Lewis; M. Randall. *Thermodynamics*, (revised by K. S. Pitzer; L. Brewer) McGraw-Hill, New York, 1961, 2nd. edn.
[4] J. E. Garrod; T. M. Herrington. *J. Chem. Educ.*, 1969, **46**, 165.

2.8 CHEMICAL POTENTIAL OF A SOLUTE IN SOLUTION

For a solution of solute j in a solvent at temperature T and pressure p, the chemical potential of substance j is related to the molality m_j (Table 2.8.1). The reference state for solute j is a solution in which both m_j and the activity coefficient γ_j are unity at pressure p and temperature T. The activity coefficient for solute j is defined such that as the solution becomes more dilute so the properties of solute j approach ideal. The activity coefficient γ_j cannot be negative or zero. In the definition of γ_j, we envisage

Table 2.8.1. Chemical potential of a solute in solution

Fixed T; fixed p; closed system = solution; solvent = substance 1;
solute = substance j.
System prepared using
(a) n_1 moles of solvent, and
(b) n_j moles of solute; molality $m_j = n_j/n_1$
Definition:

$$\mu_j(\text{sln};T;p) = \mu_j(\text{sln};T;p;m_j = 1:\gamma_j = 1) + RT\ln(m_j\gamma_j/m^0)$$

where $\text{limit}(m_j \to 0)\gamma_j = 1.0$ at all T and p.
and $\mu_j(\text{sln};T;p;m_j = 1;\gamma_j = 1)$ is the chemical potential of solute j in the reference solution; an ideal solution containing solute j where $m_j/\text{mol kg}^{-1} = 1.0$ at the same T and p.

Table 2.8.2. Volumes

$$V(\text{sln}; T; p) = n_1 V_1(\text{sln}; T; p) + n_j V_j(\text{sln}; T; p)$$

limit$(n_j \rightarrow 0)$ $V_1(\text{sln}; T; p) = V_1^*(1; T; p)$.
Then from Table 2.8.1:

$$V_j(\text{sln}; T; p) = V_j(\text{sln}; T; p; m_j = 1; \gamma_j = 1) + RT(\partial \ln(\gamma_j)/\partial p)_T$$

From the definition of γ_j

$$\text{limit}(m_j \rightarrow 0)\ V_j(\text{sln}; T; p) = V_j(\text{sln}; T; p; m_j = 1; \gamma_j = 1) = V_j^\infty(\text{sln}; T; p)$$

Also $V_j(\text{sln}; T; p^0; m_j = 1; \gamma_j = 1) = V_j^0(\text{sln}; T)$

Table 2.8.3. Enthalpies

$$H(\text{sln}; T; p) = n_1 H_1(\text{sln}; T; p) + n_j H_j(\text{sln}; T; p)$$

limit$(n_j \rightarrow 0) H_1(\text{sln}; T; p) = H_1^*(1; T; p)$, and $H_j(\text{sln}; T; p) = H_j^\infty(\text{sln}; T; p)$
From Table 2.8.1:

$$H_j(\text{sln}; T; p) = H_j(\text{sln}; T; p; m_j = 1; \gamma_j = 1) - RT^2(\partial \ln \gamma_j / \partial T)p$$

From the definition of γ_j,
limit$(m_j \rightarrow 0)$ $H_j(\text{sln}; T; p) = H_j(\text{sln}; T; p; m_j = 1; \gamma_j = 1) = H_j^\infty(\text{sln}; T; p)$.
Also $H_j(\text{sln}; T; p^0; m_j = 1; \gamma_j = 1) = H_j^0(\text{sln}; T)$

Table 2.8.4. Isobaric equilibrium heat capacities

$$C_\text{p}(\text{sln}; T; p) = n_1 C_{\text{p}1}(\text{sln}; T; p) + n_j C_{\text{p}j}(\text{sln}; T; p)$$

limit$(n_j \rightarrow 0) C_{\text{p}1}(\text{sln}; T; p) = C_{\text{p}1}^*(1; T; p)$
From Table 2.8.3, $C_{\text{p}j}(\text{sln}; T; p) = C_{\text{p}j}(\text{sln}; T; p; m_j = 1; \gamma_j = 1) - RT^2(\partial^2 \ln \gamma_j / \partial T^2)_p$
$- 2RT(\partial \ln \gamma_j / \partial T)_p$
From the definition of γ_j,
limit$(m_j \rightarrow 0) C_{\text{p}j}(\text{sln}; T; p) = C_{\text{p}j}(\text{sln}; T; p; m_j = 1; \gamma_j = 1) = C_{\text{p}j}^\infty(\text{sln}; T; p)$

a series of solutions where m_j decreases (e.g. 0.1, 10^{-2}, 10^{-3}, ...) but where the number of moles of solvent remains constant. The equation for μ_j is tested against the criteria discussed in section 2.7 by deriving equations for the dependences on composition of partial molar volume, partial molar enthalpy and partial molar isobaric heat capacity (Tables 2.8.2–2.8.4 respectively). V_j, H_j and C_{pj}, extrapolated in the limit of infinite dilution (symbol ∞) equal the corresponding partial molar properties in the reference state (Tables 2.8.2–2.8.4). Hence the criteria are satisfied, the extrapolated properties being meaningful.

Table 2.8.5. Chemical potential of solute in solution

$$(\partial \mu_j / \partial p)_T = V_j$$

Hence,

$$\mu_j(\text{sln}; T; p; m_j = 1; \gamma_j = 1) - \mu_j(\text{sln}; T; p^0; m_j = 1; \gamma_j = 1) = \int_{p^0}^{p} V_j^{\infty}(\text{sln}; T)\,\mathrm{d}p$$

Therefore, from Table 2.8.1,

$$\mu_j(\text{sln}; T; p) = \mu_j^0(\text{sln}; T) + RT\ln(m_j\gamma_j/m^0) + \int_{p^0}^{p} V_j^{\infty}(\text{sln}; T)\,\mathrm{d}p$$

Also $\mu_j(\text{sln}; T; p^0; m_j = 1; \gamma_j = 1) = \mu_j^0(\text{sln}; T)$

The partial molar volume of a solute at infinite dilution, $V_j^{\infty}(\text{sln}; T; p)$ relates the chemical potentials of a solute j in reference solutions at different pressures. The standard chemical potential of solute j in solution $\mu_j^0(\text{sln}; T)$ is defined (Table 2.8.5) in terms of an ideal solution where $m_j = 1.0$ at temperature T and pressure, p^0, in the same solvent. A clear distinction is drawn between the standard states defined here and in section 2.7.

A given liquid mixture [1] comprises bromine and tetrachloromethane where $n(Br_2) \ll n(CCl_4)$ and $p = p^0$ at temperature T. We concentrate attention on the properties of bromine. Description I treats the system as a mixture of liquid bromine and liquid CCl_4, compares the properties of bromine in the system with those of pure bromine (at T and p) and expresses the chemical potential of bromine in the mixture using the procedure outlined in section 2.7. However, this comparison may be unrewarding. In pure bromine, each molecule is surrounded by other bromine molecules whereas in the system under discussion each bromine molecule is surrounded, for the most part, by molecules of CCl_4. Description II treats the system as a solution where Br_2 is the solute and CCl_4 is the solvent. The standard state (Table 2.8.5) for bromine is as a solute in solution. In this state each bromine molecule is surrounded (effectively) by an expanse of CCl_4 molecules. The properties of bromine (e.g. vapour pressure) in the two standard states (i.e. liquid bromine and bromine in a solution in CCl_4) are quite different. The two standard states differ in another crucial respect. We can measure the properties of pure bromine but we cannot measure directly the properties of bromine as a solute in its standard state. If we prepare a solution where $m(Br_2) = 1.0$ in CCl_4 (at p^0 and T) it is unlikely that $\gamma(Br_2)$ is unity. (For this system $\gamma(Br_2)$ will be close to but not exactly unity.) The properties of Br_2 in this reference solution are obtained by extrapolation procedures based on the properties of real solutions.

The molar volume of liquid water at 298.15 K and ambient pressure is 18.07 cm^3. Densities of liquids containing water can be analysed to yield the molar volume of water at infinite dilution in a solvent, e.g. $V^{\infty}(H_2O$; solvent $= CH_3OH$; 298.15 K; ambient $p)$. These partial molar volumes differ [2] from $V^*(H_2O$; 1; 298.15 K; $p)$ and depend on solvent (Table 2.8.6). A similar example is shown in Table 2.8.7 for urea

Table 2.8.6. Partial molar volume of water

$V^{\infty}(H_2O$; 1;298 K; ambient p) as a solute	
Solvent	$V^{\infty}(H_2O$; sln)/cm^3 mol^{-1}
THF	17.00
benzene	22.55
CH_3OH	14.41
ethyl alcohol	13.81
n-propyl alcohol	15.09
n-butyl alcohol	16.18
n-pentyl alcohol	16.89
$V^*(H_2O$; 1; 298.15 K; 101325 N m^{-2})/cm^3 mol^{-1} =	18.07

Table 2.8.7. V^{∞}(urea) in six solvents at 2981.5 K and ambient pressure

Solvent	V^{∞}(urea; sln)/cm^3 mol^{-1}
H_2O	44.24
CH_3OH	36.97
C_2H_5OH	40.75
Formamide	45.34
DMF	39.97
DMSO	41.86

[3] in various solutions. Hence one can explore the dependence of V_j^{∞} for a given solute on solvent.

References to section 2.8

[1] G. N. Lewis; M. Randall. *Thermodynamics*, (revised by K. S. Pitzer; L. Brewer), McGraw-Hill, New York, 2nd edn., 1961.
[2] M. Sakurai; T. Nakagawa. *J. Chem. Thermodyn.*, 1982, **14**, 269; 1984, **16**, 171; *Bull. Chem. Soc. Japan*, 1982, **55**, 1641.
[3] D. Hamilton; R. H. Stokes. *J. Soln. Chem.*, 1972, **1**, 213.

2.9 CHEMICAL POTENTIALS AND RELATED PROPERTIES OF SOLUTES

A given solution contains n_1 moles of solvent and n_j moles of solute. At fixed temperature and pressure, the Gibbs–Duhem equation requires that, $n_1\,d\mu_1 + n_j\,d\mu_j = 0$. This equation provides a quantitative relationship between the activity coefficient of a solute and the practical osmotic coefficient (Table 2.9.1).

Table 2.9.1. Solvent and solute—Gibbs–Duhem equation

Fixed T and fixed p. A solution where composition =
m_j moles of solute in 1 kg of solvent i.e. $n_1 M_1 = 1.0\,\mathrm{kg}$
From equation (2.5.1), $(1/M_1)\,\mathrm{d}\mu_1 + m_j\,\mathrm{d}\mu_j = 0$
From method II of Table 2.7.6, $\mathrm{d}\mu_1(\mathrm{sln}; T; p) = -RTM_1\,\mathrm{d}(\phi m_j)$
From Table 2.8.1, $\mathrm{d}\mu_j = RT\,\mathrm{d}[\ln(m_j\gamma_j/m^0)]$.
Therefore, $-\mathrm{d}(\phi m_j) + m_j\,\mathrm{d}[\ln(m_j\gamma_j/m^0)] = 0$

Analysis; $-\phi\,\mathrm{d}m_j - m_j\,\mathrm{d}\phi + m_j[(\mathrm{d}m_j/m_j) + \mathrm{d}\ln\gamma_j] = 0$
$\mathrm{d}m_j - \phi\,\mathrm{d}m_j - m_j\,\mathrm{d}\phi + m_j\,\mathrm{d}\ln\gamma_j = 0$

Hence $\mathrm{d}[m_j(1-\phi)] + m_j\,\mathrm{d}\ln\gamma_j = 0$

Table 2.9.2. Partial molar entropy of a solute

$$S_j = -(\partial\mu_j/\partial T)p$$

From $\mu_j(\mathrm{sln}; T; p) = \mu_j(\mathrm{sln}; m_j = 1; \gamma_j = 1; T; p) + RT\ln(m_j/m^0) + RT\ln\gamma_j$
Hence $S_j(\mathrm{sln}; T; p) = S_j(\mathrm{sln}; m_j = 1; \gamma_j = 1; T; p) - R\ln(m_j/m^0) - R\ln\gamma_j$
$- RT(\partial\ln\gamma_j/\partial T)p$
or $S_j(\mathrm{sln}; T; p) = S_j(\mathrm{sln}; m_j = 1; \gamma_j = 1; T; p) - R\ln(m_j\gamma_j/m^0)$
$- RT(\partial\ln\gamma_j/\partial T)p$
Also, $S_j(\mathrm{sln}; m_j; \mathrm{id}; T; p) = S_j(\mathrm{sln}; m_j = 1; \gamma_j = 1; T; p) - R\ln(m_j/m^0)$
Then, $\mathrm{limit}(m_j \to 0)\ S_j(\mathrm{sln}; T; p) = -\infty$
Also $S_j(\mathrm{sln}; m_j = 1; T; p^0) = S_j^0(\mathrm{sln}; T)$
Hence $\mu_j^0(\mathrm{sln}; T) = H_j^0(\mathrm{sln}; T) - TS_j^0(\mathrm{sln}; T)$

Table 2.9.3. Thermodynamic properties of solvents

From Table 2.7.6, $\mu_1(\mathrm{sln}; T; p) = \mu_1^*(1; T; p) - \phi RTM_1 m_j$
Volumes; $V_1(\mathrm{sln}; T; p) = V_1^*(1; T; p) - RTM_1 m_j(\partial\phi/\partial p)_T$
For an ideal solution $(\partial\phi/\partial p)_T = 0$.
Hence, $V_1(\mathrm{sln}; \mathrm{id}; T; p) = V_1^*(1; T; p)$
Entropies; $S_1(\mathrm{sln}; T; p) = S_1^*(1; T; p) + \phi RM_1 m_j + RTM_1 m_j(\partial\phi/\partial T)p$
For an ideal solution, $(\partial\phi/\partial T)_p = 0.0$ and $\phi = 1.0$

$$S_1(\mathrm{sln}; \mathrm{id}; T; p) = S_1^*(1; T; p) + RM_1 m_j;$$

where limit $(m_j \to 0) S_1(\mathrm{sln}; T; p) = S_1^*(1; T; p)$
Enthalpies $\mu_1(\mathrm{sln}; T; p)/T = \mu_1^*(1; T; p)/T - \phi RM_1 m_j$

$$H_1(\mathrm{sln}; T; p) = H_1^*(1; T; p) + RT^2 M_1 m_j(\partial\phi/\partial T)p$$

For ideal solutions, $(\partial\phi/\partial T)_p = 0.0$ and $\phi = 1.0$;
and $H_1(\mathrm{sln}; \mathrm{id}; T; p) = H_1^*(1; T; p)$
limit $(m_j \to 0)$ $H_1(\mathrm{sln}; T; p) = H_1^*(1; T; p)$, and $C_{p1}(\mathrm{sln}; T; p) = C_{p1}^*(1; T; p)$

The story is completed in Table 2.9.2 with reference to the partial molar entropies of solutes. The difference $[S_j(\text{sln; id}; T; p) - S_j(\text{sln}; m_j = 1; \gamma_j = 1; T; p)]$ is a function of the molality, m_j. An analogous set of equations are derived in Table 2.9.3 for solvents.

2.10 THE QUANTITIES $((\partial H/\partial \xi)_{T;p}$, $(\partial C_p/\partial \xi)_{T;p}$ AND $(\partial V/\partial \xi)_{T;p}$

$(\partial H/\partial \xi)_{T;p}$ and $(\partial V/\partial \xi)_{T;p}$ describe the changes in enthalpy and volume, respectively, for unit change in ξ at constant temperature and pressure. We re-express these quantities in terms of the partial molar enthalpies and volumes of the chemical substances involved in chemical reaction. In Tables 2.10.1–2.10.3 we also derive equations for ideal solutions.

Table 2.10.1. Partial derivative $(\partial H/\partial \xi)_{T,p}$

Fixed temperature; fixed pressure. Closed system, n_1 moles of solvent and n_i moles of solute where $i = 2, 3, ..$

$$H = \Sigma(j = 2; j = i) n_j H_j + n_1 H_1$$

If n_1 is constant,

$$(\partial H/\partial \xi)_{T;p} = \Sigma(j = 2; j = \text{i})[n_j(\partial H_j/\partial \xi) + H_j(\partial n_j/\partial \xi)] + n_1(\partial H_1/\partial \xi)$$

From the Gibbs–Duhem equation,

$$\Sigma(j = 2; j = \text{i}) v_j(\partial H_j/\partial \xi) + n_1(\partial H_1/\partial \xi) = 0$$

Hence, $(\partial H/\partial \xi)_{T;p} = \Sigma(j = 2; j = \text{i}) v_j H_j$
But, $H_j(\text{sln}; T; p) = H_j^{\infty}(\text{sln}; T; p) - RT^2(\partial \ln \gamma_j/\partial T)_p$
Hence,

$$(\partial H/\partial \xi)_{T;p} = \Sigma(j = 1; j = \text{i}) v_j[H_j^{\infty}(\text{sln}; T; p) - RT^2(\partial \ln \gamma_j/\partial T)_p]$$

For an ideal solution. $(\partial H/\partial \xi)_{T;p} = \Sigma(j = 2; j = \text{i}) v_j H_j^{\infty}(\text{sln}; T; p)$
At $p = p^0$. $(\partial H/\partial \xi)_T = \Sigma(j = 2; j = \text{i}) v_j H_j^0(\text{sln}; T)$
From Table 2.4.8, also for an ideal solution
$C_p(A = 0) = C_p(\xi^{eq}) - (1/T)\{\partial \xi/\partial A\}_{T;p}^{eq}[\Sigma(j = 2; j = \text{i}) v_j H_j^{\infty}(\text{sln}; T)]^2$

2.11 DESCRIPTION OF SALT SOLUTIONS

If solute j is a salt, one mole of a simple salt forms, on complete dissociation, v_+ moles of cations and v_- moles of anions. (The term 'simple' means that the salt forms only two ionic substances.) The chemical potential for salt j is expressed [1] as the stoichiometrically weighted sum of the chemical potentials of cations and anions (equation 2.11.1).

Table 2.10.2. The derivative $(\partial V/\partial \xi)_{T;p}$

System (see Table 2.10.1) $V = \Sigma(j = 2; j = i) n_j V_j + n_1 V_1$
If n_1 is constant,

$$(\partial V/\partial \xi)_{T;p} = \Sigma(j = 2; j = i)[n_j(\partial V_j/\partial \xi) + V_j(\partial n_j/\partial \xi)] + n_1(\partial V_1/\partial \xi)$$

From the Gibbs–Duhem equation,

$$\Sigma(j = 2; j = i) n_j(\partial V_j/\partial \xi) + n_1(\partial V_1/\partial \xi) = 0$$

Therefore, $(\partial V/\partial \xi)_{T;p} = \Sigma(j = 2; j = i) v_j V_j$
But, $V_j(\text{sln}; T; p) = V_j^{\infty}(\text{sln}; T; p) + RT(\partial \ln \gamma_j/\partial p)_T$
Hence, $(\partial V/\partial \xi)_{T;p} = \Sigma(j = 2; j = i) v_j[V_j^{\infty}(\text{sln}; T; p) + RT(\partial \ln \gamma_j/\partial \xi)_T]$
For an ideal solution. $(\partial V/\partial \xi)_{T;p} = \Sigma(j = 2; j = i) v_j V_j^{\infty}(\text{sln}; T; p)$

Table 2.10.3. The derivative $(\partial C_p/\partial \xi)_{T;p}$

From Table 2.10.1, $(\partial C_p/\partial \xi)_{T;p} = \Sigma(j = 2; j = i) v_j C_{pj}(\text{sln}; T; p)$
For an ideal solution $(\partial C_p/\partial \xi)_{T;p} = \Sigma(j = 2; j = i) v_j C_{pj}^{\infty}(\text{sln}; T; p)$

Table 2.11.1. Mean ionic parameters

System: molality of salt $= m_j$.

$$m_{\pm} = \text{mean ionic molality } (m_{\pm})^{v} = (m_{+})^{v+}(m_{-})^{v-}$$

where m_+ and m_- are molalities of cation and anion respectively; v_+ and v_- are the stoichiometric numbers.
By definition $(\gamma_{\pm})^{v} = (\gamma_{+})^{v+}(\gamma_{-})^{v-}$; where $\gamma_{\pm}$ = mean ionic activity coefficient.
Hence, $m_{\pm} = Q m_j$, where $Q^{v} = ((v_{+})^{v+}(v_{-})^{v-})$.

$$\mu_j(\text{aq}; T; p) = v_{+}\mu_{+}(\text{aq}; T; p) + v_{-}\mu_{-}(\text{aq}; T; p) \tag{2.11.1}$$

This equation introduces the concept of ionic chemical potentials. Deviations in the properties of salt solutions from a defined ideal are attributed to the corresponding deviations from ideal of the properties of cations and anions. In these terms, single ion activity coefficients γ_+ and γ_- are defined. The properties of salt solutions are characterized by (geometric) mean ionic activity coefficients, $\gamma_{\pm}$, whereby the properties of ions in solutions are incorporated into mean ionic properties characterizing the salt (Table 2.11.1).

While as chemists we concentrate attention on the properties of ions in solution, we must draw the relevant equations together in order to describe the properties of salts. At the risk of overstating the point, experiments yield information about the properties of electrically neutral salt solutions.

Table 2.11.2. Chemical potential of salts in aqueous solutions

Solution: m_j = molality of salt; temperature T and pressure p.
By definition,

$$\mu_j(\text{aq}; T; p) = \mu_j^0(\text{aq}; T) + \nu RT \ln(Q m_j \gamma_\pm / m^0) + \int_{p^0}^{p} V_j^\infty(\text{aq}; T; p)\,\mathrm{d}p$$

where, limit $(m_j \to 0)\gamma_\pm = 1.0$ at all T and p.

Table 2.11.3. Partial molar properties of salts in aqueous solution

From Table 2.11.2

$$\mu_j(\text{aq}; m_j = 1; \gamma_\pm = 1; T; p) = \mu_j^0(\text{aq}; m_j = 1; \gamma_\pm = 1; T) + \int_{p^0}^{p} V_j^\infty(\text{aq}; T; p)\,\mathrm{d}p$$

Then, $\mu_j(\text{aq}; T; p) = \mu_j(\text{aq}; m_j = 1; \gamma_\pm = 1; T; p) + \nu RT \ln(Q m_j \gamma_\pm / m^0)$
Also $\mu_j(\text{aq}; m_j = 1; \gamma_\pm = 1; T; p^0) = \mu_j^0(\text{aq}; T)$

Table 2.1.4. Chemical potential of 1 : 1 electrolytes

In aqueous solutions; for a 1 : 1 salt, $\nu = 2$, $Q = 1$

$$\mu_j(\text{aq}; T; p) = \mu_j(\text{aq}; m_j = 1; \gamma_\pm = 1; T; p) + 2RT \ln(m_j \gamma_\pm / m^0)$$

The extrathermodynamic assumption in equation (2.11.2) is extended to the properties of salt j in its standard state in aqueous solution;

$$\mu_j^0(\text{aq}; T) = \nu_+ \mu_+^0(\text{aq}; T) + \nu_- \mu_-^0(\text{aq}; T) \tag{2.11.2}$$

If these assumptions are accepted, the chemical potential of a salt in solution at temperature T and pressure p is related to the molality m_j using the equations in Tables 2.11.2 and 2.11.3. The chemical potential of a 1 : 1 salt in solution is related to the molality m_j using the equation given in Table 2.11.4. Then $\mu_j^0(\text{aq}; T)$ is the chemical potential of salt j in an aqueous solution where $m_j/\text{mol kg}^{-1} = 1$ and the mean ionic activity coefficient, $\gamma_\pm = 1$ at temperature T and the standard pressure, p^0. $V_j^\infty(\text{aq}; T; p)$ is the limiting partial molar volume of the salt (i.e. at infinite dilution) in aqueous solution at pressure p and temperature T.

References to section 2.11

[1] R. A. Robinson; R. H. Stokes. *Electrolyte Solutions*, Butterworths, London, 1959, 2nd. edn.

2.12 PARTIAL MOLAR PROPERTIES OF SALTS AND SOLVENTS

For an aqueous salt solution containing a simple salt, the definition of the chemical potential of salt j (Table 2.11.1) satisfies the criteria discussed in section 2.7 for an equation relating the chemical potential of substance j to the composition of a solution. Consequently in the limit of infinite dilution, partial molar enthalpies and volumes of salts in solution at temperature T and pressure p equal the corresponding partial molar properties in an ideal solution (Tables 2.12.1 and 2.12.2).

If the difference between ambient and standard pressure is ignored, the reference partial molar enthapy $H_j(\text{sln}; m_j = 1; \gamma_{\pm} = 1; T; p = \text{ambient})$ equals the standard partial molar enthalpy $H_j^0(\text{sln}; T)$. The partial molar entropy of a given salt in an aqueous solution is related to the molality of salt in solution. In the limit that m_j tends to zero, the partial molar entropy tends to plus infinity (Table 2.12.3.

The chemical potential of water in a given aqueous salt solution is related to the molality of each ionic substance using a practical osmotic coefficient ϕ. In an ideal salt solution, $\phi = 1.0$. In many cases, the system under discussion contains a single salt j and hence the composition of the solution appears as the product νm_j (Table 2.12.4).

Table 2.12.1. Partial molar volumes of salts(aq)

$$V_j = (\partial \mu_j / \partial p)_T$$

From Table 2.11.2:

$$V_j(\text{aq}; T; p) = V_j(\text{aq}; m_j = 1; \gamma_{\pm} = 1; T; p) + \nu RT(\partial \ln \gamma_{\pm} / \partial p)_T$$

From the definition of $\gamma_{\pm}$;

$$\text{limit}(m_j \to 0) V_j(\text{aq}; T; p) = V_j(\text{aq}; m_j = 1; \gamma_{\pm} = 1; T; p) = V_j^{\infty}(\text{aq}; T; p)$$

Also $V_j(\text{aq}; m_j = 1; \gamma_{\pm} = 1; T; p^0) = V_j^0(\text{aq}; T)$

Table 2.12.2. Partial molar enthalpies of salts(aq)

Using $\{(\partial \mu_j / T)/\partial T\}p = -H_j/T^2$
From Table 2.11.2:

$$H_j(\text{aq}; T; p) = H_j(\text{aq}; m_j = 1; \gamma_{\pm} = 1; T; p) - \nu RT^2 \{\partial \ln \gamma_{\pm} / \partial T\}_p$$

From the definition of $\gamma_{\pm}$,

$$\text{limit}(m_j \to 0) H_j(\text{aq}; T; p) = H_j(\text{aq}; m_j = 1; \gamma_{\pm} = 1; T; p) = H_j^{\infty}(\text{aq}; T; p)$$

Also, $\text{limit}(m_j \to 0) C_{pj}(\text{aq}; T; p) = C_{pj}(\text{aq}; m_j = 1; \gamma_{\pm} = 1; T; p) = C_{pj}^{\infty}(\text{aq}; T; p)$
Further, $H_j(\text{aq}; m_j = 1; \gamma_{\pm} = 1; T; p^0) = H_j^0(\text{aq}; T)$,

$$C_{pj}(\text{aq}; m_j = 1; \gamma_{\pm} = 1; T; p^0) = C_{pj}^0(\text{aq}; T)$$

and $V_j(\text{aq}; m_j = 1; \gamma_{\pm} = 1; T; p^0) = V_j^0(\text{aq}; T)$

Table 2.12.3. Partial molar entropies of salts(aq)

$$\{(\partial \mu_j / \partial T\}_p = -S_j$$

From Table 2.11.4:

$S_j(\text{aq}; T; p) = S_j(\text{aq}; m_j = 1; \gamma_\pm = 1; T; p) - \nu R \ln(Q m_j \gamma_\pm / m^0) - \nu RT\{\partial \ln \gamma_\pm / \partial T\}_p$

From the definition of $\gamma_\pm$, $\text{limit}(m_j \to 0) S_j(\text{aq}; T; p) = +\infty$
Further, $S_j(\text{aq}; m_j = 1; \gamma_\pm = 1; T; p^0) = S_j^0(\text{aq}; T)$.

Table 2.12.4. Solvents in aqueous salt solutions

Temperature T; pressure p.
Molality of each ionic substance i (for $i = 2, 3, \ldots, k$) $= m_i$;
By definition (water = substance 1),

$$\mu_1(\text{aq}; T; p) = \mu_1^0(1; T) + \left[\int_{p^0}^{p} V_1^*(1; T; p)\,\text{d}p \right] - \phi RTM_1 \Sigma(i = 2; i = k) m_i$$

At pressure p, $\mu_1(\text{aq}; T; p) = \mu_1^*(1; T; p) - \phi RTM_1\, \Sigma(i = 2; i = k) m_i$
Example:
one salt in solution; molality $m_j; m_+ = \nu_+ m_j; m_- = \nu_- m_j$

$$\nu = \nu_+ + \nu_-$$

Then, $\mu_1(\text{aq}; T; p) = \mu_1^*(1; T; p) - \phi RTM_1 \nu m_j$.

2.13 SALTS AND SOLVATES IN AQUEOUS SOLUTION

A theme running through this chapter is a concern for descriptions of systems. The importance of description emerges in a review of the properties of solvates in solution. Specifically we are often faced with a situation where we add salt MX to water and where we are convinced by experimental evidence that the cation exists as a hydrate $M(H_2O)_h$. An important example concerns hydrogen ions in aqueous solution. Two descriptions [1] are (a) H^+(aq), and (b) H_3O^+(aq). As a starting point we assume that an aqueous solution comprises n_j moles of electrolyte HX and n_1 moles of water (Table 2.13.1).

A simple description of the solution is in terms of H^+(aq). The chemical potential of H^+(aq) describes the change δG when $\delta n(H^+)$ moles of H^+ are added at constant $n(H_2O)$, $n(X^-)$, T, and p. The chemical potential of H^+(aq) is related to the molality $m(H^+)$ $[= n(H^+)/n_1 M_1]$ and the single ion activity coefficient $\gamma(H^+)$. In a similar fashion, we define the partial molar volume, enthalpy and isobaric heat capacity for H^+(aq) (Table 2.13.2). The Gibbs function for the solution is given by the equation in Table 2.13.3. Another description [1, 2] of hydrogen ions in solution is in terms of H_3O^+(aq). The chemical potential of H_3O^+(aq) describes the change δG when $\delta n(H_3O^+)$ ions are added at constant $n(H_2O)$, $n(X^-)$, T and p (Table 2.13.4). The

Table 2.13.1. Description of system in terms of H^+(aq)

System: fixed T and pressure.
(1) n_j moles of HX. (2) n_1 moles of water. At equilibrium,

$$G^{eq}(aq; T; p) = n_1 \mu_1(H_2O; aq; T; p) + n_j \mu_j(H^+; aq; T; p) + n_j \mu_j(X^-; aq; T; p)$$

where $\mu(H^+; aq; T; p) = [\partial G/\partial n(H^+)]_{n(H_2O); n(X^-); T; p}$

$$\mu(H^+; aq; T; p) = \mu^0(H^+; aq; T) + RT \ln(m(H^+)\gamma(H^+)/m^0) + \int_{p^0}^{p} V^\infty(H^+; aq; T; p)\, dp$$

where $m(H^+) = n(H^+)/n_1 M_1$
For the 1 : 1 electrolyte H^+X^- with $\nu = 2$,

$$\mu(H^+X^-; aq; T; p) = \mu^0(H^+X^-; aq; T) + 2RT \ln(m(H^+X^-)\gamma_\pm(H^+X^-)/m^0) + \int_{p^0}^{p} V^\infty(H^+X^-; aq; T; p)\, dp$$

where $m(H^+X^-) = n(H^+X^-)/n_1 M_1$, and $\text{limit}\{m(H^+X^-) \to 0\}\gamma_\pm(H^+X^-) = 1.0$

Table 2.13.2. Partial molar properties of H^+(aq)

From Table 2.13.1, for the 1 : 1 electrolyte H^+X^-,

$$\mu(H^+X^-; aq; T; p) = \mu(m(H^+X^-) = 1; \gamma_\pm = 1; aq; T; p) + 2RT \ln(m(H^+X^-)\gamma_\pm(H^+X^-)/m^0)$$

$$V(H^+; aq; T; p) = (\partial V/\partial n(H^+))_{T; p; n_1; n(X^-)}$$

$$H(H^+; aq; T; p) = (\partial H/\partial n(H^+))_{T; p; n_1; n(X^-)}$$

For electrolyte H^+X^-,

$$V^\infty(H^+X^-; aq; T; p) = V^\infty(H^+; aq; T; p) + V^\infty(X^-; aq; T; p)$$

$$H^\infty(H^+X^-; aq; T; p) = H^\infty(H^+; aq; T; p) + H^\infty(X^-; aq; T; p)$$

and

$$C_p^\infty(H^+X^-; aq; T; p) = C_p^\infty(H^+; aq; T; p) + C_p^\infty(X^-; aq; T; p)$$

Table 2.13.3. Gibbs energy $G(\text{aq}; T; p)$

System: Table 2.13.1
Solvent:

$$\mu_1(\text{aq}; T; p) = \mu_1^*(1; T; p) - \phi RT 2m(\text{H}^+\text{X}^-)$$

Solute:

$$\mu(\text{H}^+\text{X}^-; \text{aq}; T; p) = \mu(m(\text{H}^+\text{X}^-) = 1; \gamma_\pm = 1; \text{aq}; T; p)$$
$$+ 2RT \ln\{m(\text{H}^+\text{X}^-)\gamma_\pm(\text{H}^+\text{X}^-)/m^0)$$

Solution:

$$G(\text{aq}; T; p) = n_1\mu_1(\text{aq}; T; p) + n(\text{H}^+\text{X}^-)\mu(\text{H}^+\text{X}^-; \text{aq}; T; p)$$
$$= n_1\mu_1(\text{aq}; T; p) + n(\text{H}^+)\mu(\text{H}^+; \text{aq}; T; p) + n(\text{X}^-)\mu(\text{X}^-; \text{aq}; T; p)$$

Table 2.13.4. Chemical potential $\mu(\text{H}_3\text{O}^+)$

System: Table 2.13.1. Description: solutes are H_3O^+ and X^-

$$\mu(\text{H}_3\text{O}^+; \text{aq}; T; p) = (\partial G/\partial n(\text{H}_3\text{O}^+))_{T; p; n(\text{H}_2\text{O}); n(\text{X}^-)}$$

molality $m(\text{H}_3\text{O}^+\text{X}^-) = n(\text{H}_3\text{O}^+\text{X}^-)/n(\text{H}_2\text{O})M_1; n(\text{H}_2\text{O}) = n_1 - n(\text{H}_3\text{O}^+)$

$$\mu(\text{H}_3\text{O}^+\text{X}^-; \text{aq}; T; p) = \mu(m(\text{H}_3\text{O}^+\text{X}^-) = 1; \gamma_\pm = 1; \text{aq}; T; p)$$
$$+ 2RT \ln(m(\text{H}_3\text{O}^+\text{X}^-)\gamma_\pm(\text{H}_3\text{O}^+\text{X}^-)/m^0)$$

where $\text{limit}(m(\text{H}_3\text{O}^+\text{X}^-) \to 0)\gamma_\pm(\text{H}_3\text{O}^+\text{X}^-) = 1.0$
$\text{limit}(m(\text{H}_3\text{O}^+\text{X}^-) \to 0)\{m(\text{H}_3\text{O}^+\text{X}^-)/m(\text{H}^+\text{X}^-)\} = 1.0$

chemical potential $m(\text{H}_3\text{O}^+\text{X}^-; \text{aq}; T; p)$ of the electrolyte $\text{H}_3\text{O}^+\text{X}^-$ is related to the molality $m(\text{H}_3\text{O}^+\text{X}^-)$ and the mean ionic activity coefficient $\gamma_\pm(\text{H}_3\text{O}^+\text{X}^-)$. We compare two descriptions of the same system prepared according to the conditions set out in Table 2.13.5. In description I, the system is an aqueous solution containing H^+ and X^- ions, whereas in description II the system is an aqueous solution containing H_3O^+ and X^- ions. In effect we have shifted a mole of water for each mole of H^+ ions from consideration as part of the solvent in description I to part of the solute in description II forming H_3O^+ ions. These descriptions are linked through two formulations of $G^{\text{eq}}(\text{aq}; T; p)$ which are identical for both systems (as are V^{eq}, S^{eq} and H^{eq}) (Table 2.13.5).

The equality of the total Gibbs function and equilibrium chemical potentials of substances common to both descriptions leads to an equation relating $\mu(\text{H}^+; \text{aq}; T; p)$ and $\mu(\text{H}_3\text{O}^+; \text{aq}; T; p)$. In Table 2.13.6 we take the analysis a stage further and use the

Table 2.13.5. H^+X^-(aq) and $H_3O^+X^-$(aq)

System: Table 2.13.1

Description I:

$$G(\text{aq; I}; T; p) = n_1\mu_1(\text{aq; I}; T; p) + n_j\mu(H^+; \text{aq}; T; p) + n_j\mu(X^-; \text{aq; I}; T; p)$$

Description II

$$G(\text{aq; II}; T; p) = (n_1 - n_j)\mu_1(\text{aq; II}; T; p) + n_j\mu(H_3O^+; \text{aq}; T; p) + n_j\mu(X^-; \text{II; aq}; T; p)$$

At equilibrium (a) $G(\text{I; aq}; T; p) = G(\text{II; aq}; T; p)$

(b) $\mu_1(\text{aq; I}; T; p) = \mu_1(\text{aq; II}; T; p)$

and (c) $\mu(X^-; \text{aq; I}; T; p) = \mu(X^-; \text{aq; II}; T; p)$

Hence, at equilibrium $\mu(H^+; \text{aq}; T; p) + \mu_1(\text{aq}; T; p) = \mu(H_3O^+; \text{aq}; T; p)$

Table 2.13.6. Comparison of descriptions

System: Table 2.13.1

From Table 2.13.5:

Description I: $m(H^+) = n(H^+)/n_1M_1$ $\quad m(\text{I}; X^-) = n(X^-)/n_1M_1$

Description II: $m(H_3O^+) = n(H_3O^+)/(n_1 - n_j)M_1$;

$m(\text{II}; X^-) = n(X^-)/(n_1 - n_j)M_1$

Also, $\mu^{eq}(H_3O^+; \text{aq}; T; p) = \mu^{eq}(H^+; \text{aq}; T; p) + \mu_1^{eq}(\text{aq}; T; p)$

Hence, $\mu(m(H_3O^+X^-) = 1; \gamma_\pm = 1; \text{aq}; T; p)$

$+ 2RT\ln(\gamma_\pm(H_3O^+X^-)m(H_3O^+X^-)/m^0)$

$= \mu(m(H^+X^-) = 1; \gamma_\pm = 1; H^+X^-; \text{aq}; T; p)$

$+ 2RT\ln(\gamma_\pm(H^+X^-)m(H^+X^-)/m^0) + \mu_1^*(1; T; p) - 2\phi(\text{I})RTm(H^+X^-)M_1$ (a)

But $\text{limit}(n_j \to 0)m(H^+X^-)/m(H_3O^+) = 1.0$

$$\gamma_\pm(H^+X^-)/\gamma_\pm(H_3O^+X^-) = 1.0$$

and $\phi(\text{I})m(H^+X^-) = 0$

Hence, $\mu(m(H_3O^+X^-) = 1; \gamma_\pm = 1; \text{aq}; T; p)$

$= \mu(m(H^+X^-) = 1; \gamma_\pm = 1; H^+; \text{aq}; T; p) + \mu_1^*(H_2O; 1; T; p)$ (b)

Also, $V^\infty(H_3O^+; \text{aq}; T; p) = V^\infty(H^+; \text{aq}; T; p) + V_1^*(H_2O; 1; T; p)$

and $C_p^\infty(H_3O^+; \text{aq}; T; p) = C_p^\infty(H^+; \text{aq}; T; p) + C_{p1}^*(H_2O; 1; T; p)$

equations relating chemical potential and composition for H^+(aq) and H_3O^+(aq). The difference in reference chemical potentials of these ions equals the chemical potential of water at the same T and p. In Table 2.13.7, we combine two equations in Table 2.13.6 to obtain equations relating the mean ionic activity coefficients $\gamma_\pm(H_3O^+X^-)$ and $\gamma_\pm(H^+X^-)$. Clearly in dilute aqueous solutions where $m(H^+X^-)/m(H_3O^+X^-)$ is approximately unity and $\phi(\text{I})m(H^+X^-)M_1$ is negligibly small, the two mean ionic activity coefficients are equal but this approximation becomes less acceptable with increase in the ratio $n(H^+X^-)/n(H_2O)$.

Table 2.13.7. Mean ionic activity coefficients

Equation (a); Table 2.13.6

$$\ln \gamma_{\pm}(H_3O^+X^-) = \ln \gamma_{\pm}(H^+X^-) + \ln \{m(H^+X^-)/m(H_3O^+X^-)\}$$
$$- \phi(I)m(H^+X^-)M_1 + [0.5RT]\{\mu(m(H^+X^-) = 1; \gamma_{\pm} = 1; aq; T;)$$
$$+ \mu_1^*(1; T; p) - \mu(m(H_3O^+X^-) = 1; \gamma_{\pm} = 1; aq; T; p)\}$$

Using equation (b) of Table 2.213.6;

$$\ln \gamma_{\pm}(H_3O^+X^-) = \ln \gamma_{\pm}(H^+X^-)$$
$$+ \ln \{m(H^+X^-)/m(H_3O^+X^-)\} - \phi(I)m(H^+X^-)M_1$$

and in dilute solutions,

$$\ln (\gamma_{\pm}(H_3O^+X^-)) = \ln (\gamma_{\pm}(H^+X^-))$$

References to section 2.13

[1] M. Eigen. *Angew. Chem. Internat. edn.*, 1964, **3**, 1.
[2] H. L. Clever. *J. Chem. Educ.*, 1963, **40**, 637.

3

Chemical equilibria in solution

3.1 INTRODUCTION

In this chapter we are concerned with chemical equilibria [1, 2] in solutions and their description using equilibrium constants and related parameters. For the most part we concentrate attention on chemical equilibria. However, many treatments of the kinetics of chemical reactions in solution are based on transition state theory [3–5]. This theory assumes that initial and transition states in a solution are in chemical equilibrium. Hence a patina of thermodynamic theory is used in understanding dependences of rate constants on temperature and pressure.

Chemical analysis of a given solution at pressure p and temperature T reveals that the system comprises two neutral solutes, substances X and Y in chemical equilibrium (Table 3.1.1). Hence $\mu_x^{eq}(\mathrm{sln}; T; p)$ is the equilibrium chemical potential of solute X at temperature T, pressure p and in the presence of all other substances. The molalities of solutes X and Y at equilibrium are m_x^{eq} and m_y^{eq} respectively. At equilibrium, the chemical potentials of solute X and solute Y are equal; $\mu_x^{eq}(\mathrm{sln}; T; p) = \mu_y^{eq}(\mathrm{sln}; T; p)$ (Table 2.1.3). These equilibrium chemical potentials are related to the equilibrium molalities of both substances X and Y and their activity coefficients, γ_x and γ_y (Table 3.1.1). The standard reaction Gibbs energy, $\Delta_r G^0(T)$, equals the difference between the standard chemical potentials of reactants and products (Table 3.1.1). $\Delta_r G^0(T)$ is the increase in G at T and p when ξ advances by unity under standard conditions. A dimensionless quantity $K^0(T)$ is the standard equilibrium constant, defined at standard pressure p^0 (Table 3.1.1). $K^0(T)$ is related to the equilibrium composition of the system at temperature T and pressure p. By definition $K^0(T)$ is independent of pressure and composition but dependent on temperature. The chemical equilibrium in Table 3.1.1 is symmetric where the stoichiometric sum $(\Sigma(j = 1; j = \mathrm{i})\nu_j)$ is zero. A general treatment is given in Table 3.1.2 for chemical equilibria involving i solutes. (Another treatment of chemical equilibria is based on volume fractions [6])

Thermodynamic parameters $\Delta_r G^0(T)$ and $K^0(T)$ are the parents of a family of equilibrium thermodynamic parameters (Table 3.1.3). The standard reaction enthalpy $\Delta_r H^0(T)$ at temperature T is related to the differential dependence of $\ln K^0$ on

Table 3.1.1. Chemical equilibria

Temperature T; pressure p.
Description: aqueous solution of solutes X and Y; water = solvent 1.
Chemical equilibrium: $X \rightleftharpoons Y$, where $\nu_x = -1$, and $\nu_y = +1$.
At equilibrium, $A^{eq} = 0$, G^{eq} = minimum. Hence,

$$-\mu_x^{eq}(\text{sln}; T; p) + \mu_y^{eq}(\text{aq}; T; p) = 0$$

$$\mu_x^0(\text{sln}; T) + RT\ln(m_x^{eq}\gamma_x^{eq}/m^0) + \int_{p^0}^{p} V_x^{\infty}(\text{sln}; T; p)\mathrm{d}p$$

$$= \mu_y^0(\text{sln}; T) + RT\ln(m_y^{eq}\gamma_y^{eq}/m^0) + \int_{p^0}^{p} V_y^{\infty}(\text{sln}; T; p)\,\mathrm{d}p$$

By definition, $\Delta G^0(\text{sln}; T) = -RT\ln K^0(\text{sln}; T) = \mu_y^0(\text{sln}; T) - \mu_x^0(\text{sln}; T)$

Hence, $K^0(T) = [m_y^{eq}\gamma_y^{eq}/m_x^{eq}\gamma_x^{eq}]\exp\left[\int_{p^0}^{p} \{\Delta_r V^{\infty}(\text{sln}; T; p)/RT\}\,\mathrm{d}p\right]$

where $\Delta_r V^{\infty}(\text{sln}; T; p) = V_y^{\infty}(\text{sln}; T; p) - V_x^{\infty}(\text{sln}; T; p)$
Assumptions: (a) $p = p^0$; $K^0(T) = [m_y^{eq}\gamma_y^{eq}/m_x^{eq}\gamma_x^{eq}]$
(b) the solution is ideal; $K^0(T) = m_y^{eq}/m_x^{eq}$

Table 3.1.2. Chemical equilibria—general statement

Fixed T and p.
For a chemical equilibrium involving i solutes in solution,

$$\Sigma(j = 1; j = i)\ \nu_j\mu_j^{eq}(\text{system}; T; p) = 0$$

By definition, $\Delta_r G^0(T) = -RT\ln K^0(T) = \Sigma(j = 1; j = i)\ \nu_j\mu_j^0(\text{sln}; T)$

$$K^0(T) = \left\{\prod(j = 1; j = i)(m_j^{eq}\gamma_j^{eq}/m^0)^{\nu(j)}\right\}\exp\left[\int_{p^0}^{p} (\Delta_r V^{\infty}(T; \text{sln})/RT)\,\mathrm{d}p\right]$$

$$\Delta_r V^{\infty}(\text{sln}; T; p) = \Sigma(j = 1; j = i)\ \nu_j V_j^{\infty}(\text{sln}; T; p)$$

temperature. The standard reaction isobaric heat capacity, $\Delta_r C_p^0$ describes the differential dependence of $\Delta_r H^0$ on T. The standard reaction entropy $\Delta_r S^0$ is related to $\Delta_r G^0$ and $\Delta_r H^0$ (Table 3.1.3). The equilibrium composition of a given system depends on pressure and so we introduce parameters which characterize equilibria at temperature T and pressure p. The chemical potential of solute j is related to the molality m_j using the reference chemical potential $\mu_j(\text{sln}; m_j = 1; \gamma_j = 1; T; p)$. An equilibrium constant, $K(\text{sln}; T; p)$, is related to standard equilibrium constant $K^0(\text{sln}; T)$ through $\Delta_r V^{\infty}(\text{sln}; T; p)$ (Table 3.1.4). In other words, $K(\text{sln}; T; p)$ is related to $K^0(\text{sln}; T)$ through the partial molar volumes at infinite dilution of solutes involved in a chemical

Table 3.1.3. Standard quantities for chemical equilibria

Standard equilibrium constant $K^0(T)$ at temperature T;

$$\Delta_r G^0(T) = -RT\ln K^0(T) = \Sigma(j=1;j=\text{i})\ \nu_j\mu_j^0(T) = \Delta_r H^0(T) - T\Delta_r S^0(T)$$

Enthalpy: $\Delta_r H^0 = -T^2[\mathrm{d}(\Delta_r G^0/T)/\mathrm{d}T] = \Sigma(j=1;j=\text{i})\ \nu_j H_j^0$
van't Hoff equation*: $\Delta_r H^0 = RT^2[\mathrm{d}\ln K^0/\mathrm{d}T] = -R[\mathrm{d}\ln K^0/\mathrm{d}(1/T)]$
Isobaric heat capacity: $\Delta_r C_p^0 = (\mathrm{d}\Delta_r H^0/\mathrm{d}T) = \Sigma(j=1;j=\text{i})\nu_j C_{pj}^0$

$$\Delta_r H^0(T_2) - \Delta_r H^0(T_1) = \int_{T_1}^{T_2} \Delta_r C_p^0\,\mathrm{d}T$$

$$\Delta_r S^0 = R\ln K^0 + RT(\mathrm{d}\ln K^0/\mathrm{d}T) \text{ and, } \Delta_r C_p^0 = \mathrm{d}\Delta_r S^0/\mathrm{d}\ln T$$

Hence, $\Delta_r S^0(T_2) - \Delta_r S^0(T_1) = \int_{T_1}^{T_2} \Delta_r C_p^0\,\mathrm{d}\ln T$

* $\mathrm{d}T^{-1} = -T^2\,\mathrm{d}T$, or $\mathrm{d}T = -T^2\,\mathrm{d}T^{-1}$

Table 3.1.4. Equilibrium constant, $K(\mathrm{sln}; T; p)$

$$K(\mathrm{sln}; T; p) = \Pi(j=1;j=\text{i})[(m_j^{\mathrm{eq}}\gamma_j^{\mathrm{eq}}/m^0)^{\nu(j)}]$$

$$-RT\ln K(\mathrm{sln}; T; p) = \Sigma(j=1;j=\text{i})\ \nu_j\mu_j(\mathrm{sln}; m_j=1; \gamma_j=1; T; p)$$

At temperature T,

$$K(\mathrm{sln}; p) = K^0(\mathrm{sln}; T)\exp\left[\int_{p^0}^{p}\{-\Delta_r V^\infty(\mathrm{sln})/RT\}\,\mathrm{d}p\right]$$

where, by definition, $K^0(\mathrm{sln}; T)$ is independent of pressure
Also, $K^0(\mathrm{sln}; T) = K(\mathrm{sln}; T; p^0)$

equilibrium. The differential dependence of $\ln K(\mathrm{sln}; T; p)$ on pressure at constant temperature is related to $\Delta_r V^\infty(\mathrm{sln}; T; p)$ and on temperature at constant pressure to $\Delta_r H^\infty(\mathrm{sln}; T; p)$ (Table 3.1.5). Both $\Delta_r H^\infty(\mathrm{sln}; T; p)$ and $\Delta_r V^\infty(\mathrm{sln}; T; p)$ depend on T and p. But through the equation described in Table 2.6.2, the dependence of $\Delta_r H^\infty(\mathrm{sln}; T; p)$ on pressure is related to $\Delta_r V^\infty(\mathrm{sln}; T; p)$ (Table 3.1.5). At this stage, the relationship cannot be taken further because the integral term cannot be evaluated. Standard equilibrium constants $K^0(\mathrm{sln}; T)$ are related to standard enthalpies and entropies of reaction (Table 3.1.6).

References for section 3.1

[1] M. L. McGlashan. *Chemical Thermodynamics*, Academic Press, London, 1979.
[2] G. N. Lewis; M. Randall. *Thermodynamics*, (revised by K. S. Pitzer; L. Brewer) McGraw-Hill, New York, 1961, 2nd. edn.

Table 3.1.5. Dependence of ln $K(\text{sln}; T; p)$ on temperature and pressure

From Table 3.1.4, at fixed T, $[\partial \ln K(\text{sln})/\partial p]_T = -\Delta_r V^\infty(\text{sln})/RT$.
From Table 3.1.4,

$$\mathrm{d} \ln K(\text{sln}; P)/\mathrm{d}T = [\mathrm{d} \ln K^0(\text{sln})/\mathrm{d}T] - (\mathrm{d}/\mathrm{d}T)\left[\int_{p^0}^{p} [\Delta_r V^\infty(\text{sln})/RT]\, \mathrm{d}p\right]$$

Hence (Table 3.1.3)

$$\mathrm{d} \ln K(\text{sln}; p)/\mathrm{d}T = (1/RT^2)\left[\Delta_r H^0(\text{sln}; T) - T^2(\mathrm{d}/\mathrm{d}T)\left[\int_{p^0}^{p} [\Delta_r V^\infty(\text{sln})/T]\, \mathrm{d}p\right]\right]$$

$$= (1/RT^2)\left[\Delta_r H^0(\text{sln}; T) + \left[\int_{p^0}^{p} [\Delta_r V^\infty(\text{sln}) - T(\mathrm{d}/\mathrm{d}T)\Delta_r V^\infty(\text{sln})/T]\, \mathrm{d}p]\right]\right]$$

From Table 2.6.2, at fixed temperature T,

$$\Delta_r H^\infty(\text{sln}; p) - \Delta_r H^0(\text{sln}) = \int_{p^0}^{p} \{\Delta_r V^\infty(\text{sln}) - T(\mathrm{d}/\mathrm{d}T)\Delta_r V^\infty(\text{sln})\}\, \mathrm{d}p$$

Hence, $[\partial \ln K(\text{sln})/\partial T]_p = \Delta_r H^\infty(\text{sln})/RT^2$
where, $\Delta_r H^\infty(\text{sln}; T; p) = \Sigma(j = 1; j = \text{i})\ v_j H_j^\infty(\text{sln}; T; p)$
Also, $(\partial \Delta_r H^\infty(\text{sln})/\partial T)_p = \Delta_r C_p^\infty(\text{sln}; T; p)$
If $p = p^0$, $K^0(\text{sln}; T) = K(\text{sln}; T; p^0)$, $\Delta_r H^0(\text{sln}; T) = \Delta_r H^\infty(\text{sln}; T; p^0)$
and $\Delta_r C_\text{p}^0(\text{sln}; T) = \Delta_r C_\text{p}^\infty(\text{sln}; T; p^0)$

Table 3.1.6. Standard equilibrium constants

For chemical equilibria involving i solutes at temperature T,

$$\Delta_r G^0(\text{sln}; T) = \Sigma(j = 1; j = \text{i}) v_j \mu_j^0(\text{sln}; T) = -RT \ln K^0(\text{sln}; T)$$

$$= \Delta_r H^0(\text{sln}; T) - T\Delta_r S^0(\text{sln}; T)$$

[3] S. Glasstone; K. J. Laidler; H. Eyring. *The Theory of Rate Processes*, McGraw-Hill, New York, 1941.
[4] K. J. Laidler. *J. Chem. Educ.*, 1988, **65**, 540.
[5] K. J. Laidler; M. C. King. *J. Phys. Chem.*, 1983, **97**, 2657.
[6] P. G. Wright. *J. Chem. Soc. Faraday Trans. 1*, 1986, **82**, 2565.

3.2 RATE CONSTANTS AND DERIVED PARAMETERS

A given solution contains a number of solutes at temperature T and pressure p where the solvent is liquid, 1_l. As a result of chemical reaction, the composition depends on time. In describing the kinetics of reactions in solution, almost universal practice uses

Table 3.2.1. Transition state theory

First order unimolecular reactions. Fixed T and fixed p.
Chemical reaction: reaction molecule $j \rightleftharpoons \neq \rightarrow$ products.
Description of system: reactant j and transition state (symbol: $\neq$) are solutes in solvent, liquid substance 1.
Hence for reactant j,

$$c_j = n_j/V,\ m_j = n_j/n_1 M_1,\ c_{\neq} = n_{\neq}/V \text{ and } m_{\neq} = n_{\neq}/n_1 M_1$$

Then, $c_{\neq}/c_j = m_{\neq}/m_j$
Equilibrium:

$$^{\neq}K^0 = \{(m_{\neq}\gamma_{\neq}/m^0)/(m_j\gamma_j/m^0)\} \exp\left[\int_{p^0}^{p} [+\Delta^{\neq} V^{\infty}(\text{sln}; T)/RT]\,dp\right]$$

where, $\Delta^{\neq} G^0 = -RT \ln {}^{\neq}K^0(T) = \mu_{\neq}^0(\text{sln}; T) - \mu_j^0(\text{sln}; T)$

Hence, $c_{\neq} = \{c_j\gamma_j/\gamma_{\neq}\}\, {}^{\neq}K^0 \exp\left[\int_{p^0}^{p} [-\Delta^{\neq} V^{\infty}(\text{sln}; T)/RT]\,dp\right]$

But for reactant j, $-dc_j/dt = kc_j = (k_B T/h)\kappa c_{\neq}$
Therefore, at temperature T

$$k = \kappa[k_B T/h]^{\neq}K^0(T)\{\gamma_j/\gamma_{\neq}\} \exp\left[\int_{p^0}^{p} [-\Delta_r V^{\infty}(\text{sln})/RT]\,dp\right]$$

For ideal solution at pressure p, $k(\text{id}: T) = \kappa(k_B T/h)K(\text{sln}; T)$.
If $k^{\#} = k(\text{id}; T)s^0$, then* $k^{\#}(T) = \kappa[k_B T s^0/h]^{\neq}K(\text{sln}; T)$

*Units: $[1] = [1][\text{J K}^{-1}\text{ K s/J s}][1]$; $s^0 = 1$ s
k_B = Boltzmann constant, h = Planck constant, κ = transmission coefficient.

concentrations. Experiment yields the dependences of c_j on time for substance j; $c_j = n_j/V$ where V is the volume of the system at time t.

Transition state theory uses the equation given in Table 3.2.1 for a first order unimolecular chemical reaction in aqueous solution. The outcome (Tables 3.2.2) is an equation relating rate constants to standard molar enthalpies of activation $\Delta^{\neq} H^0$ and standard molar entropies of activation $\Delta^{\neq} S^0$. Standard equilibrium constants $^{\neq}K^0$ are defined by analogy with standard equilibrium constants. The important difference is that one term has been removed in the partition function for the transition state corresponding to translation along the reaction coordinate.

A similar analysis is presented in Tables 3.2.3 and 3.2.4 except that conversion from concentration to molality introduces the density of the solution. In Tables 3.2.1 and 3.2.3 we define a dimensionless kinetic parameter $k^{\#}$, using s^0 to identify the unit of time, one second. Rate constant k refers to the properties of a system at time t. Activity coefficients for reactants and transition state characterize the system at one instant during the course of reaction. Because the composition of the system changes

Table 3.2.2. First order rate constants and transition state theory

Chemical reaction: X(aq) $\rightarrow$ Y(aq);
Assumptions: (a) transmission coefficient, $\kappa = 1$
(b) ideal solutions: $\gamma_x = \gamma_{\neq} = 1$; $k = k(\text{id})s^0$

$$k = (k_B T s^0/h)\, K^0(\text{aq}; T) \exp\left[\int_{p^0}^{p} [-\Delta^{\neq} V^{\infty}(\text{sln}; T; p)/RT]\, dp\right]$$

or $k = (k_B T s^0/h)\, {}^{\neq}K(\text{aq}; T; p)$
Then, $-RT \ln {}^{\neq}K(\text{aq}; T; p) = \Delta^{\neq} H^{\infty}(\text{aq}; T; p) - T\Delta^{\neq} S(\text{aq}; m_j = 1; \gamma_j = 1; T; p)$
$\Delta^{\neq} G^0(\text{sln}; T) = -RT \ln {}^{\neq}K^0(\text{aq}; T) = \Delta^{\neq} H^0(\text{aq}; T) - T\Delta^{\neq} S^0(\text{aq}; T)$
where $\Delta^{\neq} H^{\infty}(\text{aq}; T; p) = \Delta^{\neq} H(\text{aq}; m_j = 1; \gamma_j = 1; T; p)$
Also, $\Delta^{\neq} C_p^{\infty}(\text{aq}; T; p) = \Delta^{\neq} C_p(\text{aq}; m_j = 1; \gamma_j = 1; T; p)$.

with time these activity coefficients are time-dependent and hence so is the rate constant k. Only for systems where these coefficients are independent of composition is k independent of time. Properties ${}^{\neq}K^0$ and $\Delta^{\neq} G^0$ are defined with reference to standard states of solutes in solution. Therefore the density of the solution and the quantity m^0 preserve the dimensional integrity of the equations for rate constant k.

3.3 CHEMICAL EQUILIBRIA INVOLVING SOLVENT

In the preceding sections, the solvent was treated as chemically passive and did not enter directly the stoichiometric equation for chemical equilibria. However, for many systems, this is not the case. An example of this class of equilibrium is the dissociation of a weak acid in aqueous solution. Here we take up the theme established in Chapter 2 where we were concerned with various descriptions of hydrogen ions in aqueous solution. A given system is prepared by adding $n^0(\text{HA})$ moles of weak acid HA to $n^0(H_2O)$ moles of water. In Table 3.3.1, we explore the equilibrium state in terms of a description of hydrogen ions as H^+(aq). In the second description, hydrogen ions in aqueous solution are written as H_3O^+ (Table 3.3.2). Then $K^0(\text{II}; T)$ describes the difference between the standard chemical potentials of H_3O^+, A^- and HA in aqueous solution and of liquid water. However, we showed (Chapter 2) that $\mu^0(H^+; \text{aq}; T)$ and $\mu^0(H_3O^+; \text{aq}; T)$ are related (Table 2.13.5). Although $K^0(\text{I}; T)$ and $K^0(\text{II}; T)$ emerge from different descriptions, it turns out that they are equal (Table 3.3.3). Further $\Delta_r H^0(\text{I}; T) = \Delta_r H^0(\text{II}; T)$.

Similarly there are at least two descriptions for the ionisation of water. In Table 3.3.4 the self-dissociation is expressed in terms of the formation of H^+ and OH^- ions. In a second description of the ionization (Table 3.3.5), hydrogen ions in solution are present as H_3O^+ ions. The two treatments come together (Table 3.3.6).

Table 3.2.3. Second order rate constants and transition state theory

Solutes X and Y in aqueous solution: e.g. $c_x = n_x/V$; $m_x = n_x/n_1 M_1$.
Chemical reaction: X(aq) + Y(aq) → products.
Assumption (a) transmission coefficient, $\kappa = 1$

$$\text{rate constant } k = (k_B T/h)\,{}^{\neq}K^0(\text{aq}; T)\{\gamma_x\gamma_y/\gamma_{\neq}\}\{1/m^0\rho_1^*(1; T; p)\}$$

$$\times \exp\left[\int_{p^0}^{p}[-\Delta^{\neq}V^{\infty}(\text{sln}; T)/RT]\,dp\right]$$

$$\Delta^{\neq}V^{\infty}(\text{aq}; T; p) = V_{\neq}^{\infty}(\text{aq}; T; p) - V_x^{\infty}(\text{aq}; T; p) - V_y^{\infty}(\text{aq}; T; p)$$

Assumptions: ideal solution. Hence,

$$\text{rate constant, } k = (k_B T/h)^{\neq}K^0(\text{aq}; T)[1/m^0\rho_1^*(1; T; p)]$$

$$\times \exp\left[\int_{p^0}^{p}[-\Delta^{\neq}V^{\infty}(\text{sln}; T)/RT]dp\right]$$

If $k^{\#} = c^0 s^0 k$, at pressure p then

$$k^{\#} = (k_B T c^0 s^0/h)\,{}^{\neq}K(\text{aq}; T; p)/m^0\rho_1^*(1; T; p)$$

where, $c^0 = 1.0\ \text{mol m}^{-3}$ and $s^0 = 1\ \text{s}$

$$\Delta^{\neq}G^0(\text{aq}; T) = -RT\ln{}^{\neq}K^0(\text{aq}; T)$$

$$= \mu_{\neq}^0(\text{aq}; T) - \mu_x^0(\text{aq}; T) - \mu_y^0(\text{aq}; T)$$

At pressure p,

$$\Delta^{\neq}G(\text{sln}; m_j = 1; \gamma_j = 1; T) = \Delta^{\neq}H^{\infty}(\text{sln}; T) - T\Delta^{\neq}S(\text{sln}; m_j = 1; \gamma_j = 1; T; p)$$

where $\Delta^{\neq}H^{\infty}(\text{sln}; T; p) = \Delta^{\neq}H(\text{sln}; m_j = 1; \gamma_j = 1; T; p)$.
Also, $\Delta^{\neq}C_p^{\infty}(\text{sln}; T; p) = \Delta^{\neq}C_p(\text{sln}; m_j = 1; \gamma_j = 1; T; p)$

Table 3.2.4. Activation parameters

$$\ln\{k^{\#}(\text{id})K/T\} = \ln\{k_B s^0/h\} + \ln{}^{\neq}K(\text{sln}; T; p)$$

$$(\partial \ln k^{\#}/\partial p)_T = (\partial \ln{}^{\neq}K/\partial p)_T = -\Delta^{\neq}V^{\infty}/RT$$

$$\{\partial \ln(k^{\#}/T)/\partial T\}_p = (\partial \ln{}^{\neq}K/\partial T)_p = \Delta^{\neq}H^{\infty}/RT^2$$

Table 3.3.1. Description 1: weak acid in aqueous solution

Equilibrium: $HA(aq) \rightleftharpoons H^+(aq) + A^-(aq)$
$n^{eq}(HA) \quad n^{eq}(H^+) \quad n^{eq}(A^-)$

With substance 1 being H_2O: $m^{eq}(HA) = n^{eq}(HA)/n_1^0 M_1$;
$m^{eq}(H^+) = n^{eq}(H^+)/n_1^0 M_1$

$$m^{eq}(A^-) = n^{eq}(A^-)/n_1^0 M_1$$

At equilibrium, $\mu^{eq}(HA; aq; T; p) = \mu^{eq}(H^+; aq; T; p) + \mu^{eq}(A^-; aq; T; p)$
or $\mu^{eq}(HA; aq; T; p) = \mu^{eq}(H^+A^-; aq; T; p)$ where H^+A^- is a 1 : 1 electrolyte.
Then $\Delta_r G^0(I; T) = -RT\ln K^0(I; T) = \mu^0(H^+A^-; aq; T) - \mu^0(HA; aq; T)$
where $\Delta_r V^\infty(I; aq; T; p) = V^\infty(H^+A^-; aq; T; p) - V^\infty(HA; aq; T; p)$

and $K^0(I; T) = [m^{eq}(H^+A^-)\gamma_\pm^{eq}(H^+A^-)]^2/[m^0 m^{eq}(HA)\gamma^{eq}(HA)]$

$$\times \exp\left[\int_{p^0}^{p} \{\Delta_r V^\infty(I; aq; T)/RT\}\, dp\right]$$

Assumptions: (a) $dp = 0$; (b) $\gamma^{eq}(HA) = 1.0$

$$K^0(1; T) = \{m^{eq}(H^+A^-)\gamma_\pm^{eq}(H^+A^-)\}^2/m^0\, m^{eq}(HA)$$

Table 3.3.2. Weak acid in aqueous solution—description II

System: Table 3.3.1. Substance 1 = H_2O
Chemical equilibrium: $HA(aq) + H_2O(aq) \rightleftharpoons H_3O^+(aq) + A^-(aq)$
$n^{eq}(HA) \quad n_1^{eq} \quad n^{eq}(H_3O^+) \quad n^{eq}(A^-)$

$$m^{eq}(HA) = n^{eq}(HA)/[n_1^0 - n^{eq}(H_3O^+)]M_1$$

$$m^{eq}(H_3O^+) = n^{eq}(H_3O^+)/[n_1^0 - n^{eq}(H_3O^+)]M_1$$

At equilibrium; $\mu^{eq}(HA; aq; T; p) + \mu^{eq}(H_2O; aq; T; p) = \mu^{eq}(H_3O^+A^-; aq; T; p)$
where $H_3O^+A^-$ is a 1 : 1 electrolyte.
Then, $\Delta_r G^0(II; T) = -RT\ln K^0(II; T)$
$= \mu^0(H_3O^+A^-; aq; T) - \mu^0(HA; aq; T) - \mu^0(H_2O; 1; T)$
$\Delta_r V^\infty(II; aq; T; p) = V^\infty(H_3O^+A^-; aq; T; p) - V^\infty(HA; aq; T; p) - V_1^*(1; T; p)$
where $K^0(II; T) = [m^{eq}(H_3O^+A^-)\gamma_\pm^{eq}(H_3O^+A^-)]^2/[m^0\, m^{eq}(HA)\gamma^{eq}(HA)]$

$$\times \exp[M_1\varphi^{eq}\{2m^{eq}(H_3O^+A^-) + m^{eq}(HA)\}]\exp\left[\int_{p^0}^{p} \{\Delta_r V^\infty(II; aq; T; p)/RT\}\, dp\right]$$

Assumptions; (a) $dp = 0$, (b) $\varphi^{eq}(HA) = 1$,
(c) dilute solution; $\exp[M_1\varphi^{eq}\{2m^{eq}(H_3O^+A^-) + m^{eq}(HA)\}] = 1$
Then, $K^0(II; T) = \{m^{eq}(H_3O^+A^-)\gamma_\pm^{eq}(H_3O^+A^-)\}^2/m^0\, m^{eq}(HA)$

Table 3.3.3. Weak acids in aqueous solution—descriptions I and II

From Tables 3.3.1 and 3.3.2

$$\Delta_r G^0(\text{I};T) = -RT\ln K^0(\text{I};T)$$

$$= \mu^0(\text{H}^+;\text{aq};T) + \mu^0(\text{A}^-;\text{aq};T) - \mu^0(\text{HA};\text{aq};T)$$

$$\Delta_r G^0(\text{II};T) = -RT\ln K^0(\text{II};T)$$

$$= \mu^0(\text{H}_3\text{O}^+;\text{aq};T) - \mu^0(\text{A}^-;\text{aq};T) - \mu^0(\text{HA};\text{aq};T) - \mu_1^0(1;T)$$

But (Table 2.13.5) $\mu^0(\text{H}^+;\text{aq};T) = \mu^0(\text{H}_3\text{O}^+;\text{aq};T) - \mu_1^0(1;T)$
Therefore, $\Delta_r G^0(\text{I};T) = -RT\ln K^0(\text{I};T)$
$= -RT\ln K^0(\text{II};T) = \Delta_r G^0(\text{II};T)$
Similarly, $\Delta_r V^\infty(\text{I};\text{aq};T;p) = \Delta_r V^\infty(\text{II};\text{aq};T;p)$

Table 3.3.4. Description 1

Equilibrium: $\text{H}_2\text{O(aq)} \rightleftharpoons \text{H}^+\text{(aq)} + \text{OH}^-\text{(aq)}$
Composition: n_1^{eq} $\quad n^{eq}(\text{H}^+\text{OH}^-)$

where H^+OH^- is a 1 : 1 electrolyte
Thermodynamic condition: $\mu_1^{eq}(\text{aq};T;p) = \mu^{eq}(\text{H}^+\text{OH}^-;\text{aq};T;p)$

Then, $\mu_1^0(1;T) - 2\varphi RTM_1 m^{eq}(\text{H}^+\text{OH}^-) + \int_{p^0}^{p} V_1^*(1;T)\text{d}p$

$$= \mu^0(\text{H}^+\text{OH}^-;\text{aq};T) + 2RT\ln\{m^{eq}(\text{H}^+\text{OH}^-)\gamma_\pm^{eq}(\text{H}^+\text{OH}^-)/m^0\}$$

$$+ \int_{p^0}^{p} V^\infty(\text{H}^+\text{OH}^-;\text{aq};T)\,\text{d}p$$

where $\Delta_r G^0(\text{I};T) = -RT\ln :K^0(\text{I};T) = \mu^0(\text{H}^+\text{OH}^-;\text{aq};T) - \mu_1^0(1;T)$
$\Delta_r V^\infty(\text{I};\text{aq};T;p) = V^\infty(\text{H}^+\text{OH}^-;\text{aq};T;p) - V_1^*(1;T;p)$
and $K^0(\text{I};T) = [m^{eq}(\text{H}^+\text{OH}^-)\gamma_\pm^{eq}(\text{H}^+\text{OH}^-)/m^0]^2 \exp[2\varphi M_1 m^{eq}(\text{H}^+\text{OH}^-)]$

$$\times \exp\left[\int_{p^0}^{p} [\Delta_r V^\infty(1;\text{aq};T)/RT]\,\text{d}p\right]$$

Assumptions: (a) $\text{d}p = 0$; (b) $\exp[2\varphi M_1\, m^{eq}(\text{H}^+\text{OH}^-)] = 1$
Then, $K^0(\text{I};T) = [m^{eq}(\text{H}^+\text{OH}^-)\gamma_\pm^{eq}(\text{H}^+\text{OH}^-)/m^0]^2$

Table 3.3.5. Description II

System: Table 3.6.1.
Chemical equilibrium: $2H_2O \rightleftharpoons H_3O^+(aq) + OH^-(aq)$
$n_1^{eq} \qquad n^{eq}(H_3O^+OH^-)$

Thermodynamic condition: $2\mu_1^{eq}(aq; T; p) = \mu^{eq}(H_3O^+OH^-; aq; T; p)$
Then,

$$2\{\mu_1^0(1; T) - 2\varphi RTM_1 m^{eq}(H_3O^+OH^-) + \int_{p^0}^{p} V_1^*(1; T; p)\,dp\}$$

$$= \mu^0(H_3O^+OH^-; aq; T) + 2RT\ln\{m^{eq}(H_3O^+OH^-)\gamma_{\pm}^{eq}(H_3O^+OH^-)/m^0\}$$

$$+ \int_{p^0}^{p} V^{\infty}(H_3O^+OH^-; aq; T; p)\,dp$$

$$K^0(II; T) = [m^{eq}(H_3O^+OH^-)\gamma_{\pm}^{eq}(H_3O^+OH^-)/m^0]^2 \exp[4\varphi^{eq} M_1 m^{eq}(H_3O^+OH^-)]$$

$$\times \exp\left[\int_{p^0}^{p} \{\Delta_r V^{\infty}(sln; T; p)/RT\}\,dp\right]$$

where $\Delta_r G^0(II; T) = -RT\ln K^0(II; T)$
$= \mu^0(H_3O^+OH^-; aq; T) - 2\mu_1^0(1; T)$
and $\Delta_r V^{\infty}(II; aq; T; p) = V^{\infty}(H_3O^+OH^-; aq; T; p) - 2V_1^*(1; T; p)$
Approximations: (a) dilute solution, and (b) $p \simeq p^0$.

$$K^0(II; T) = [m^{eq}(H_3O^+OH^-)\gamma_{\pm}^{eq}(H_3O^+OH^-)/m^0]^2$$

Table 3.3.6. Comparison of descriptions

From Table 2.13.5

$$\mu^0(H_3O^+OH^-; aq; T) = \mu^0(H_3O^+; aq; T) + \mu^0(OH^-; aq; T)$$

$$= \mu^0(H^+; aq; T) + \mu^0(OH^-; aq; T) + \mu_1^0(1; T)$$

Similarly, $V^{\infty}(H_3O^+OH^-; aq; T; p) = V^{\infty}(H^+OH^-; aq; T; p) + V_1^*(1; T; p)$
Then,
$\Delta_r G^0(II; T) = \mu^0(H_3O^+OH^-; aq; T) - 2\mu_1^0(1; T)$
$= \mu^0(H^+; aq; T) + \mu^0(OH^-; aq; T) + \mu_1^0(1; T) - 2\mu_1^0(1; T)$
$= \Delta_r G^0(I; T)$
and $K^0(II; T) = K^0(I; T) = K_w^0$ (by definition)

Table 3.4.1. Self-dissociation in an aqueous binary mixture

Temperature T; pressure $\simeq p^0$
Solvent = binary liquid mixture water (liquid 1) + liquid 2

$$x_1 = n_1/(n_1 + n_2) = 1.0 - x_2$$

For solute j, m_j = number of moles of solute j in 1 kg of the mixture. Then for water (liquid 1)

$$H_2O(sln; x_2) \rightleftharpoons H^+(sln; x_2) + OH^-(sln; x_2)$$

At equilibrium; $\mu_1^{eq}(sln; x_2) = \mu^{eq}(H^+OH^-; sln; x_2)$.
Then

$$\mu_1^*(1) + RT\ln(x_1 f_1) = \mu^0(H^+OH^-; sln; x_2) + 2RT\ln\{m(H^+OH^-)\gamma_\pm/m^0\}_{x_2}^{eq}$$

where

$$\text{limit}\{m(H^+OH^-) \rightarrow \text{zero})\gamma_\pm = 1.0 \text{ at all } T \text{ and } p \text{ and at fixed } x_2$$

Then $\Delta_r G^0(\text{self dissoc}; x_2) = \mu^0(H^+OH^-; sln; x_2) - \mu_1^0(l)$ (1)
$= -2RT\ln\{m(H^+OH^-)\gamma_\pm/m^0\}_{x_2}^{eq} + RT\ln(x_1 f_1)^{eq}$

We omit the 'eq' superscript.
Then, $K_w(sln; x_2) = \{m(H^+OH^-)\gamma_\pm/m^0\}^2/(x_1 f_1)$

3.4 SELF-DISSOCIATION IN BINARY AQUEOUS MIXTURES

There is a further complexity concerning the self-dissociation of water which we should consider. We have in mind the impact of adding a second liquid (liquid 2) to water. Consequently the extent of self-dissociation of water changes. The question arises as to how we might describe this change. Any description we offer involves an *a priori* definition [1–4] of the reference states for water and for the electrolyte H^+OH^-. For water the reference state is pure water, where the chemical potential at pressure p and temperature T is $\mu_1^*(1; T; p)$. The reference state for the electrolyte is an ideal solution of unit molality in H^+OH^- in a solvent formed by a mixture of liquid 1 and liquid 2 at defined mole fraction (Table 3.4.1).

In the case of the dissociation of an acid, general formula HA, the dependence of acid dissociation constant on mixture composition at constant temperature and standard (approximately ambient) pressure (see [5, 6]) reflects the dependences on mixture composition of the standard chemical potentials of the undissociated and dissociated acid, i.e. $\mu^0(HA; sln; x_2; T)$ and $\mu^0(H^+A^-; sln; x_2; T)$.

References to section 3.4

[1] R. A. Matheson. *J. Phys. Chem.*, 1969, **73**, 3635.
[2] F. Lenzi; J. Sangster. *Can. J. Chem. Eng.*, 1973, **51**, 792.
[3] IUPAC Commission. *Pure Appl. Chem.*, 1987, **59**, 1693.
[4] M. J. Blandamer; J. Burgess; B. Clark; P. P. Duce; A. W. Hakin; N. Gosal; S. Radulovic; P. Guardado; F. Sanchez; C. D. Hubbard; E.-E. Abu Gharib. *J. Chem. Soc Faraday Trans.*, 1988, **82**, 1471.
[5] K-S. Siow; K-P. Ang. *J. Soln. Chem.*, 1989, **18**, 937.
[6] S. Goldman; P. Sagner; R. Bates. *J. Phys. Chem.*, 1971, **75**, 826.

4

Dependence of equilibrium and rate constants on pressure

According to our starting hypothesis, there exists for a given chemical equilibrium in solution a unique $K(\text{sln}; T; p)$ for defined temperature and pressure. In other words, the dependent variable $\ln K(\text{sln})$ is defined by the independent variables, T and p (Table 4.1.1). Here $K(\text{sln})$, T and p are intensive variables. Using the van't Hoff equations (Table 4.1.1), we express $\mathrm{d}\ln K(\text{sln})$ as a function of limiting molar enthalpies and volumes of reaction [1–3] under defined conditions.

Thermodynamics does not define *a priori* how $K(\text{sln})$ depends on pressure. But the pressure derivative of $\ln K(\text{sln})$ is related to $\Delta_r V^\infty(\text{sln}; T; p)$ (Table 4.1.2). $\Delta_r V^\infty(\text{sln}; T; p)$ depends on temperature and pressure. The latter dependence is described [2] in terms of a compression parameter, $\Delta_r \chi^\infty$ (Table 4.1.3).

References to section 4.1.1

[1] S. D. Hamann. *Rev. Phys. Chem. Japan*, 1980, **50**, 147.
[2] S. D. Hamann. *J. Soln. Chem.*, 1982, **11**, 63.
[3] S. D. Hamann. *High Temp–High Pressures*, 1983, **15**, 511.

4.2 TRUE AND ESTIMATED $K(T;p)$

Experimental data report the dependence of the composition of a solution on pressure at fixed temperature. These data are analysed taking account of the dependence on pressure of activity coefficients to yield the corresponding dependence of $K(T)$ on pressure at fixed temperature. We assume that there exists for defined temperature and pressure a true $K(T;p)$; i.e. $K(\text{true}; \text{sln}; T; p)$. Although it is only possible to measure p and T to finite precision, in practice quoted T and p are assumed true and so without error. Therefore experimental data comprise estimates $K(\text{sln})$ at error-free T and p.

Setting the experimental data aside, we assert there is a pattern to the dependence of $K(\text{true}; \text{sln})$ on pressure at fixed temperature. In other words, for the same system, $K(\text{true}; \text{sln}; T; p_1)$ is related to $K(\text{true}; \text{sln}; T; p_2)$. In the next stage we set down a possible relationship between $K(\text{true}; \text{sln}; T)$ and pressure and then examine how

Table 4.1.1. Dependence of K(sln) on temperature and pressure

System: chemical equilibrium; $\ln K = \ln K[T; p]$

$$\mathrm{d}\ln K(\mathrm{sln}) = [\partial \ln K(\mathrm{sln})/\partial p]_T\,\mathrm{d}p + [\partial \ln K(\mathrm{sln})/\partial T]_p\,\mathrm{d}T$$

Therefore, $\mathrm{d}\ln K(\mathrm{sln}) = -\{\Delta_r V^{\infty}(T)/RT\}\,\mathrm{d}p + \{\Delta_r H^{\infty}(p)/RT^2\}\,\mathrm{d}T$

Table 4.1.2. Dependence of $K(\mathrm{sln}; T; p)$ on pressure

Closed system; fixed temperature T.

From Table 4.1.1, $\ln K^0(\mathrm{sln}) - \ln K(\mathrm{sln}; p) = \int_p^{p^0} [-\Delta_r V^{\infty}(\mathrm{sln})/RT]\,\mathrm{d}p$

where $\ln K^0(T) \equiv \ln K(T; p^0)$
van't Hoff Equation. $\Delta_r V^{\infty}(\mathrm{sln}; T) = -RT[\partial \ln K(\mathrm{sln})/\partial p]_T$
If $p = p^0$, $\Delta_r V^{\infty}(\mathrm{sln}; T; p^0) = \Delta_r V^0(\mathrm{sln}; T) = \Sigma(j = 1; j = \mathrm{i})\nu_j V_j^0(\mathrm{sln}; T)$
where, $\Delta_r V^{\infty}(\mathrm{sln}; T; p) = \Sigma(j = 1; j = \mathrm{i})\nu_j V_j^{\infty}(\mathrm{sln}; T; p)$

Table 4.1.3. Limiting volume of reaction and related parameters

$$\Delta_r V^{\infty}(\mathrm{sln}; T) = -RT(\partial \ln K/\partial p)_T$$

and* $\Delta_r \chi^{\infty} = -(\partial \Delta_r V^{\infty}(\mathrm{sln}; T)/\partial p)_T = RT(\partial^2 \ln K/\partial p^2)_T$
or $(\partial \ln K/\partial p)_T = -\Delta_r V^{\infty}(\mathrm{sln}; T)/RT$ and $(\partial^2 \ln K/\partial p^2)_T = \Delta_r \chi^{\infty}/RT$

* Units: $\Delta_r \chi^{\infty} = [\mathrm{m^3\,mol^{-1}}]/[\mathrm{Pa}] = [\mathrm{m^3\,mol^{-1}\,Pa^{-1}}]$

satisfactorily it describes the observed dependence of $K(\mathrm{sln}; T)$ on pressure. The van't Hoff equation prompts us to express $\ln K(\mathrm{sln}; T)$ rather than $K(\mathrm{sln}; T)$ itself as a function of pressure.

Several reasons can be advanced for embarking on this exercise. At one level, we seek to identify a pattern to the dependence of $K(\mathrm{sln}; T)$ on pressure. The input to the analysis comprises $\ln K(\mathrm{sln}; T)$ at ambient and higher pressures, the range being dictated by available technology. Having identified a pattern to the data, it is then possible to interpolate the information and obtain $\ln K(\mathrm{sln}; T)$ with some degree of confidence at a selected pressure. Where the pattern is used to predict $\ln K(\mathrm{sln}; T)$ at pressures beyond the upper limit of the measured pressure range, confidence in $\ln K(\mathrm{calc}; \mathrm{sln}; T)$ rapidly falls with increase in pressure.

These arithmetic exercises are overshadowed by attempts to determine thermodynamic parameters such as limiting volumes of reaction [1–3] from observed dependences of $\ln K(\mathrm{sln}; T)$ on pressure. In order to determine, for example,

$\Delta_r V^\infty(\text{sln}; T; p)$ for a particular process, we require the partial derivative $(\partial \ln K/\partial p)_T$. In general terms we anticipate that this derivative and hence $\Delta_r V^\infty(\text{sln}; T)$ also depends on pressure. Hence we require from the analysis a quantitative description of this dependence, i.e. $\Delta_r \chi^\infty$ and $\Delta_r V^\infty$ for a given process at temperature T and at pressure p. In some cases we may be able to compare $\Delta_r V^\infty(\pi)$ at pressure π determined from the dependence of $\ln K(\text{sln}; T)$ on pressure with $\Delta_r V^\infty(\pi)$ determined directly using a dilatometer [4, 5].

References to section 4.2

[1] B. B. Owen; S. R. Brinkley. *Chem. Revs.*, 1941, **29**, 461.

[2] S. D. Hamann. *Modern Aspects of Electrochemistry* (ed. B. E. Conway; J. O'M. Bockris), Plenum Press, New York, 1974, **9**, 47.

[3] S. D. Hamann. *High Temperatures–High Pressures*, 1983, **15**, 511.

[4] S. Scrivastava: M. J. DeCicco; E. Kuo; W. J. le Noble. *J. Soln. Chem.*, 1984, **13**, 663.

[5] Y. Kitamura; T. Itoh. *J. Soln. Chem.*, 1987, **16,** 715.

4.3 DEPENDENCE OF RATE CONSTANT ON PRESSURE AT FIXED TEMPERATURE

Rate constants for chemical reaction depend on temperature and pressure (Table 4.3.1). In terms of transition state theory (Chapter 3), the differential dependence of $\ln k^{\#}$ on pressure (Table 3.2.4) is related to a volume of activation, $\Delta^{\neq} V^\infty$. While $\Delta^{\neq} G^0$ is positive consistent with the idea of activation, $\Delta^{\neq} V^\infty$ can be positive, negative, or zero. Where $\Delta^{\neq} V^\infty > 0$, the rate constant decreases with increase in pressure. Interest in volumes of activation stems from their conceptual simplicity. Further by combining $\Delta^{\neq} V^\infty$ with $\Delta_r V^\infty$ for the overall reaction considerable insight into the process of reaction often emerges [1–13].

From a practical standpoint we start out with the statement that the kinetics of reaction in a given system are characterized by true $k^{\#}$ (i.e. $\mathbf{k}^{\#}$) at true T and p. Experimental data yield estimates of k at estimated T and p. Moreover we assert there is a pattern to the underlying dependences of $\mathbf{k}^{\#}$ on true p and true T. The measured dependence of k on T and on p is used to formulate the true dependence. In other words, we assume the phenomenological rate constant for reaction at T_1 and p_1 is related to the rate constant at T_2 and p_2.

Table 4.3.1. Dependence of rate constants on pressure

From Table 3.2.4, with all derivatives at constant temperature

$$(\partial \ln k^{\#}/\partial p) = -\Delta^{\neq} V^\infty/RT$$

where for a first order unimolecular reaction,

$$\Delta^{\neq} V^\infty(\text{sln}; T; p) = V^\infty_{\neq}(\text{sln}; T; p) - V_j^\infty(\text{sln}; T; p)$$

A variety of units are used to record pressures. If the aim is to calculate a volume of activation, expressed in cubic metres per mole, there is merit in using the SI unit of pressure, the newton per square metre (or pascal). Many equations described in the literature use 'atmosphere' as the unit of pressure and a reference pressure of 1 atm. The accompanying algebra is often simplified but at some cost, especially where the integer '1' occurs in equations. Often this integer has an associated unit of pressure.

References to section 4.3

[1] S. D. Hamann. *Physico-Chemical Effects of Pressure*, Butterworths, London, 1957.
[2] T. Asano; W. J. le Noble. *Chem. Revs.*, 1978, **78**, 407.
[3] W. J. le Noble. *Progr. Phys. Org. Chem.*, 1967, **5**, 207.
[4] W. J. le Noble. *J. Chem. Educ.*, 1967, **44,** 729.
[5] W. J. le Noble. *Chem. Weekblad*, 1967, **63**, 39.
[6] W. J. le Noble. *Rev. Phys. Soc. Japan*, 1980, **50**, 207.
[7] R. van Eldik; T. Asano; W. J. Le Noble. *Chem. Revs.*, 1989, **89**, 549.
[8] E. Whalley. *Berichte Bunsenges Physik. Chemie*, 1966, **70**, 958.
[9] R. van Eldik; H. Kelm. *Rev. Phys. Soc. Japan*, 1980, **50**, 185.
[10] D. A. Palmer, H. Kelm. *Coord. Chem. Revs.*, 1981, **36**, 89.
[11] B. S. El'yanov; E. M. Vasylvitskaya. *Rev. Phys. Chem. Japan*, 1980, **50**, 169.
[12] E. Whalley. *Adv. Phys. Org. Chem.*, 1964, **2**, 93.
[13] J. Orszagh; M. Barigand; J.-J. Tondeur. *Bull. Chem. Soc. France*, 1976, 1684.

4.4 GENERAL EQUATIONS

The input to an analysis comprises $\ln K(\text{sln}; T)$ at a number of pressures. Then at temperature T,

$$\int_{p^0}^{p} \mathrm{d} \ln K = -(1/RT) \int_{p^0}^{p} \{\Delta_r V^\infty(\text{sln})\}\, \mathrm{d}p \tag{4.4.1}$$

The volume of reaction depends on pressure (at temperature T);

$$\Delta_r V^\infty(\text{sln}; p_2) \quad = \quad \Delta_r V^\infty(\text{sln}; p_1)\, \mathrm{f}(p_2/p_1) \tag{4.4.2}$$

Hence, $\ln K(\text{sln}; p) =$

$$\ln K^0(\text{sln}) - \{\Delta_r V^0(\text{sln})/RT\} \int_{p^0}^{p} f(p/p^0)\, \mathrm{d}p \tag{4.4.3}$$

Table 4.4.1. Taylor expansion

At fixed temperature

$$\ln K(\text{sln}; p) = \ln K(\text{sln}; p) + [\partial \ln K/\partial p]_\pi (p - \pi) + (1/2)[\partial^2 \ln K/\partial p^2]_\pi (p - \pi)^2 + (1/6)[\partial^3 \ln K/\partial p^3]_\pi (p - \pi)^3 + \ldots$$

Hence, $\ln K(\text{sln}; p) = \ln K(\text{sln}; p) + \{-\Delta_r V^\infty(\text{sln}; p)(p - \pi)/RT\} + \{(1/2)\Delta_r \chi^\infty(\text{sln}; \pi)(p - \pi)^2/RT\} + \{(1/6)[\partial \Delta_r \chi^\infty(\text{sln})/\partial p]_\pi (p - \pi)^3/RT\}$

To make further progress we need to identify an explicit form for the function $f(p/p^0)$ otherwise the integral cannot be evaluated. In one approach, we express the dependence at temperature T of $\ln K(\text{sln}; p)$ on pressure using a Taylor expansion about $\ln K(\text{sln}; \pi)$ at reference pressure π (Table 4.4.1). The multiplier of $[(p - \pi)/RT]$ is the limiting volume of reaction, and the multiplier of $[(p - \pi)^2/2RT]$ is $\Delta_r\chi^\infty$ at pressure π.

4.5 VOLUME OF REACTION—INDEPENDENT OF PRESSURE

In many cases the volume of reaction is independent of pressure at temperature T (Table 4.5.1). The final equation describes a linear dependence of $\ln K(\text{true})$ on $(p - \pi)/RT$ at temperature T where the intercept at $p = \pi$ is $\ln K(\text{true}; \pi)$ and the gradient yields the volume of reaction. In the analysis of the experimental data, we use the measured dependence of $\ln K$ on $(p - \pi)/RT$ to obtain estimates of $\ln K(\text{true}; \pi)$ and the limiting volume of reaction $\Delta_r V^\infty(\text{sln}; T)$. A linear least squares treatment yields the overall standard error and standard errors on these estimates. One advantage of fitting data to the independent variable $(p - \pi)/RT$ rather than $p - \pi$ is that the statistical information obtained from the least squares analysis refers directly to $\Delta_r V^\infty(\text{sln}; T)$. For many applications, π is ambient pressure such that $p - \pi$ is the gauge or 'excess' pressure. There is often merit in setting π equal to the mean of the experimental pressures so that $(p - \pi)/\pi$ spans zero. We develop this point below in more detail. At this stage we cite one example [1] which involves the dependence of $\ln K(\text{sln})$ on pressure for the first acid dissociation constant of $H_2S(aq)$ at 298.2 K. In common with many other weak acids in aqueous solution [2, 3], the acid dissociation constant increases with increase in pressure, yielding a negative limiting volume of reaction, $\Delta_r V^\infty(\text{sln})$ at 298.2 K and ambient pressure (Table 4.5.2).

In conjunction with kinetic data we use the equations in Table 4.5.3 to define the dimensionless kinetic parameter k. In this way we can be confident of not 'losing dimensions' when the equations are differentiated. Here we assume the volume of activation is independent of pressure over the measured pressure range including the reference pressure, π (Table 4.5.4). Hence $\ln k(p)$ is a linear function [4–8] of $p - \pi$, the intercept where $p - \pi$ is zero yielding $\ln k(\pi)$. The procedures are illustrated using the dependence on pressure of the rate constants describing the alkaline hydrolysis of iron(II) complex cations $[Fe(sb)_3]^{2+}$ in aqueous solutions [6]. (The input data com-

---

Table 4.5.1. Volume of reaction—independent of pressure

At temperature T:
From Table 4.4.1, if $\Delta_r\chi^\infty = 0$, then $\Delta_r V^\infty(\text{sln}) = \text{constant}$

$$d\ln K = -[\Delta_r V^\infty(\text{sln})/RT]\,dp$$

Hence*, $\ln K(p) = \ln K(\pi) - \{\Delta_r V^\infty(\text{sln})(p - \pi)/RT\}$
General form: $\ln K(p) = a_1 + a_2[(p - \pi)/RT]$

*Units: $[1] = [1] - [\text{m}^3\,\text{mol}^{-1}][\text{N}\,\text{m}^{-2}]/[\text{J}\,\text{K}^{-1}\,\text{mol}^{-1}][\text{K}]$

Table 4.5.2. First acid dissociation constant for $H_2S(aq)$

Data from [1] $T/K = 298.2$
Pressure range: $1.01 \times 10^5 \leqslant p/Pa \leqslant 2.02 \times 10^8$
Equation, Table 4.5.1, $\pi/Pa = 1.01 \times 10^5$

$$\ln K(\mathrm{sln}) = a_1 + a_2[(p - \pi)/\mathrm{Pa}][\mathrm{J\,mol^{-1}}/RT]$$

With reference pressure, $\pi = 1.01325 \times 10^5$ Pa; st error $= 4.501 \times 10^{-2}$

$$a_1 = -15.648 \pm 0.035; \quad a_2 = -12.52 \pm 0.7\,\mathrm{cm^3\,mol^{-1}}$$

$\Delta^{\neq} V^{\infty}/\mathrm{cm^3\,mol^{-1}} = -12.52 \pm 0.70$ $\ln K(\pi) = -15.6484 \pm 0.0348$

Table 4.5.3. Volumetric parameters

At temperature T
From Table 4.3.1, $(\partial \ln k^{\#}/\partial p) = -\Delta^{\neq} V^{\infty}(\mathrm{sln};p)/RT$
or* $\Delta^{\neq} V^{\infty}(\mathrm{sln};p) = -RT(\partial \ln k^{\#}/\partial p)_T$

*Units: $\Delta^{\neq} V^{\infty} = [\mathrm{J\,mol^{-1}\,K^{-1}}][\mathrm{K}][\mathrm{N\,m^{-2}}]^{-1} = [\mathrm{m^3\,mol^{-1}}]$

Table 4.5.4. Linear dependence—kinetics

At temperature T
Assumption: $(\partial \Delta^{\neq} V^{\infty}/\partial p) = 0$
Then, $\ln k^{\#}(p) = \ln k^{\#}(\pi) - \{\Delta^{\neq} V^{\infty}(\mathrm{sln})/RT\}(p - \pi)$
or $\ln k^{\#} = a_1 + a_2(p - \pi)/\mathrm{Pa}$
where $\ln k^{\#}(\pi) = a_1$ and* $\Delta^{\neq} V^{\infty}(\mathrm{sln}) = -a_2 RT/\mathrm{Pa}$

*Units: $[\mathrm{m^3\,mol^{-1}}] = [1]\,[\mathrm{J\,mol^{-1}\,K^{-1}}][\mathrm{K}]/[\mathrm{N\,m^{-2}}]$

Table 4.5.5. Alkaline hydrolysis of $[Fe(sb)_3]^{2+}$ cations in aqueous solutions

Data from [6]: $T/K = 298.2$; $1.0 \leqslant p/\mathrm{bar} \geqslant 1.34 \times 10^3$
Equation: $\ln k^{\#} = a_1 + a_2(p - \pi)/\pi$
$\pi/Pa = 1.013 \times 10^5$, $a_1 = -4.428 \pm 0.196$, $a_2 = (4.31 \pm 2.11) \times 10^{-4}$

$$\Delta^{\neq} V^{\infty}(\mathrm{aq})/\mathrm{cm^3\,mol^{-1}} = 10.5 \pm 0.5$$

sb ≡ Schiff base complex: [1,8-bis((2-pyridylmethylene)amino)-
3,6-diazooctane] iron(II)

prise ratios of measured first order rate constants, $k(p)/k(\pi)$, as a function of pressure for reactions in solutions where $[OH^-] \gg$ [complex].) For this reaction, rate constants at 298.2 K decrease with increase in pressure consistent with a positive volume of activation (Table 4.5.5).

References to section 4.5

[1] A. J. Ellis; D. W. Anderson. J. Chem. Soc., 1961, 4678.

[2] S. D. Hamann. *Modern Aspects of Electrochemistry* (ed. B. E. Conway; J. O' M. Bockris), Plenum Press, New York, 1974, **9**, 47.

[3] E. J. King. *J. Phys. Chem.*, 1969, **73**, 1220.

[4] W. J. le Noble; B. L. Yates; A. W. Scaplehorn. *J. Amer. Chem. Soc.*, 1967, **89**, 3751.

[5] R. A. Grieger, C. A. Eckert, *J. Amer. Chem. Soc.*, 1970, **92**, 7149.

[6] J. Burgess; C. D. Hubbard. *J. Amer. Chem. Soc.*, 1984, **106**, 1717.

[7] B. Anderson, F. Gronland. J. Olson, *Acta Chem. Scand.*, 1969, **23**, 2458.

[8] H. Doine; K. Ishihara; H. R. Krouse; T. W. Swaddle, *Inorg. Chem.*, 1987, **26**, 3240.

4.6 QUADRATIC DEPENDENCE

A linear dependence of $\ln K(\text{sln}; T)$ on pressure means that the volume of reaction is independent of pressure. This assumption is usually valid for a limited pressure range. In the next development we assume that the volume of reaction depends on pressure although the partial derivative $(\partial \Delta_r V^\infty/\partial p)_T$ is independent of pressure at temperature T. The argument is developed in Table 4.6.1 where we use two integrations between the pressure limits p and π. The outcome is an equation (Table 4.6.2) describing the dependence of $\ln K(\text{sln}; T; p)$ on pressure in terms of the limiting volume of reaction

Table 4.6.1. Quadratic dependence

From Table 4.1.3, $\Delta_r \chi^\infty = -(\partial \Delta_r V^\infty/\partial p)_T$

At constant temperature, $\int_\pi^p \mathrm{d}\Delta_r V^\infty = -\int_\pi^p \Delta_r \chi^\infty \, \mathrm{d}p$

Condition; $\Delta_r \chi^\infty$ = independent of pressure

$$\Delta_r V^\infty(\text{sln}; p) - \Delta_r V^\infty(\text{sln}; \pi) = -\Delta_r \chi^\infty (p - \pi)$$

Further at constant temperature,

$$\int_\pi^p \mathrm{d} \ln K(\text{sln}) = -\int_\pi^p \{\Delta_r V^\infty/RT\} \, \mathrm{d}p = \int_\pi^p [\{-\Delta_r V^\infty(\pi) + \Delta_r \chi^\infty (p - \pi)\}/RT] \, \mathrm{d}p$$

Then* $\ln K(\text{sln}; p) = \ln K(\text{sln}; \pi) - \{\Delta_r V^\infty(\text{sln}; \pi)\}[(p - \pi)/RT]$
$\quad + \Delta_r \chi^\infty [(p - \pi)^2/2RT]$

*Units: $[(1/2)\Delta_r \chi^\infty/RT](p - \pi)^2 = [1]\,[\text{m}^3\,\text{mol}^{-1}\,\text{N}^{-1}\,\text{m}^2][\text{J}\,\text{mol}^{-1}\,\text{K}^{-1}\,\text{K}]^{-1}\,[\text{N}\,\text{m}^{-2}]^2$
$= [\text{N}^{-1}\,\text{m}^5][\text{J}^{-1}][\text{N}^2\,\text{m}^{-4}] = [1]$

Table 4.6.2. Quadratic dependence in $[(p - \pi)/\pi]$

At fixed temperature:
Equation: $\ln K(p) = a_1 + a_2[(p - \pi)/\pi] + a_3[(p - \pi)/\pi]^2$

$$\text{limit}\,(p \to \pi) \ln K(p) = \ln K(\pi) = a_1$$

$$[\mathrm{d} \ln K/\mathrm{d}p] = (a_2/\pi) + 2a_3(p - \pi)/\pi^2 = -\Delta^{\neq} V^{\infty}/RT$$

$$\Delta^{\neq} V^{\infty}(p) = -RT[(a_2/\pi) + 2a_3(p - \pi)/\pi^2]; \Delta^{\neq} V^{\infty}(\pi) = -RT(a_2/\pi)$$

$$[\mathrm{d}\Delta^{\neq} V^{\infty}/\mathrm{d}p] = -RT[2a_3/\pi^2] = -\Delta_r \chi^{\infty} \text{ or } \Delta_r \chi^{\infty} = 2RTa_3/\pi^2$$

Table 4.6.3. Ionization of boric acid(aq)

$$B(OH)_3(aq) + H_2O(aq) \rightleftharpoons B(OH)_4^-(aq) + H^+(aq)$$

Data: references [2–4] $T/\mathrm{K} = 298.2$
Equation: $\ln K = a_1 + a_2\{(p - \pi)/\pi\} + a_3\{(p - \pi)/\pi\}^2$
where $1.0 \leqslant p/10^5\,\mathrm{Pa} \leqslant 10^3$ $K(298.2\,\mathrm{K};\ 101\,325\,\mathrm{Pa}) = 5.8 \times 10^{-10}$
Analysis: $\pi = 101\,325\,\mathrm{Pa}$, standard error $= 1.84 \times 10^{-4}$

$$a_1 = -21.26777 \pm (1.4 \times 10^{-4}), \quad a_2 = 1.4483 \times 10^{-3} \pm (6.53 \times 10^{-7})$$

$$a_3 = -1.45678 \times 10^{-7} \pm (6.29 \times 10^{-10})$$

$$\Delta_r V^{\infty}(298.2\,\mathrm{K}; \pi)/\mathrm{m^3\,mol^{-1}} = -35.9 \pm (1.67 \times 10^{-3})$$

$$\Delta_r \chi^{\infty}/\mathrm{m^3\,mol^{-1}\,Pa^{-1}} = -3.61 \times 10^{-4} \pm (1.56 \times 10^{-6})$$

at pressure π and the compression function $\Delta_r X^{\infty}$. The leading term is $\ln K$ at reference pressure π. The final equation corresponds to the Owen–Brinkley 'low pressure' equation [1]. Application of this equation to experimental data is illustrated using the dependence on pressure of the ionization of boric acid(aq) [2–4] at 298.15 K (Table 4.6.3). The equation is written in the form shown in Table 4.6.2 whereby a linear least squares analysis yields $\ln K(\pi)$, $\Delta_r V^{\infty}(\mathrm{sln}; T; \pi)$ and $\Delta_r \chi^{\infty}$ at temperature T. The analysis using ambient pressure as the reference pressure π leads to estimates of these three parameters, their standard errors and other statistical information.

The corresponding analysis of the dependence of rate constants on pressure is summarized in Table 4.6.4. The assumptions are that $\Delta^{\neq}\chi^{\infty}$ is independent of pressure and that the limiting volume of activation is a linear function of $p - \pi$ where π is the reference pressure (Table 4.6.5). The validity of a least squares analysis is confirmed in Table 4.6.6, where we write the equation for the true quantity $\ln k$ in terms of the true variables a_1, a_2 and a_3. We assume that the a_i parameters are independent and hence the equation for $\ln k^{\#}$ can be differentiated with respect to, say, a_1 at constant a_2 and a_3. This differential is independent of a_2 and a_3 but simply dependent on p and reference pressure π. The basic equation in Table 4.6.5 is therefore a simple quadratic

Table 4.6.4. Chemical kinetics

At temperature T.

$$\Delta^{\neq} V^{\infty}(\text{sln}; p) = \Delta^{\neq} V^{\infty}(\text{sln}; \pi) - \Delta^{\neq} \chi^{\infty}(p - \pi)$$

and $\mathrm{d}\ln k^{\#} = -\{\Delta^{\neq} V^{\infty}/RT\}\,\mathrm{d}p$

or $\int_{\pi}^{p} \mathrm{d}\ln k^{\#} = \int_{\pi}^{p} [\{-\Delta^{\neq} V^{\infty}(\text{sln}; \pi)/RT\} + \{\Delta^{\neq} \chi^{\infty}(p - \pi)/RT\}]\,\mathrm{d}p$

Hence, $\ln k^{\#}(p) = \ln k^{\#}(\pi) - \{\Delta^{\neq} V^{\infty}(\text{sln}; p)\}[(p - \pi)/RT] + [\Delta^{\neq} \chi^{\infty}/2RT](p - \pi)^2$

Table 4.6.5. Chemical kinetics—analysis

At temperature T.
From Table 4.6.4, $\ln k^{\#} = a_1 + a_2(p - \pi)/\pi + a_3((p - \pi)/\pi)^2$
$(\partial \ln k^{\#}/\partial p) = a_2/\pi + 2a_3(p - \pi)/\pi^2$ $\quad$ $(\partial^2 \ln k^{\#}/\partial p^2) = 2a_3/\pi^2$
Therefore, $\Delta^{\neq} V^{\infty}(p) = -RT[(a_2/\pi) + (2.0a_3(p - \pi)/\pi^2)]$

$$\Delta^{\neq} V^{\infty}(\pi) = -RTa_2/\pi$$

$\partial \Delta^{\neq} V^{\infty}/\partial p = -2RTa_3/\pi^2$ and $\Delta^{\neq} \chi^{\infty} = 2RTa_3/\pi^2$

Table 4.6.6. Least squares condition

From Table 4.6.5, $\ln k^{\#} = a_1 + a_2(p - \pi)/\pi + a_3((p - \pi)/\pi)^2$
At pressure p, $(\partial \ln k^{\#}/\partial a_1)_{a_i \neq 1} = 0$ $\quad$ $(\partial \ln k^{\#}/\partial a_2)_{a_i \neq 2} = (p - \pi)/\text{Pa}$
Generally, $(\partial \ln k^{\#}/\partial a_j)_{a_i \neq j} = \{(p - \pi)/\text{Pa}\}^{j-1}$ for $j = 1, 2, 3\ldots$

Table 4.6.7. Hydrolysis of benzyl chloride

Data taken from [5] (see also [5–9])

$$PhCH_2Cl(\text{sln}) + H_2O(\text{sln}) \rightarrow PhCH_2OH(\text{sln}) + H^+(\text{sln}) + Cl^-(\text{sln})$$

Temperature/K = 303.15, Solvent = water
Pressure: $1.0 \leqslant p/10^5\,\text{Pa} \leqslant 6895.0$
Reference pressure $\pi/\text{Pa} = 10^5$, st error $= 7.29 \times 10^{-3}$

$$a_1 = -(10.594 \pm 0.0044), \quad a_2 = (3.3493 \pm 0.0300) \times 10^{-4},$$

$$a_3 = -(1.5392 \pm 0.043) \times 10^{-8}$$

$\Delta^{\neq} V^{\infty}(\text{sln}; \pi)/\text{m}^3\,\text{mol}^{-1} = -8.44 \times 10^{-6}$
$\Delta^{\neq} \chi^{\infty}/\text{m}^3\,\text{mol}^{-1}\,\text{Pa}^{-1} = -7.758 \times 10^{-15}$

Table 4.6.8. Gauge pressure

From Table 4.6.6 with π = ambient pressure, π_a and $p - \pi_a = p_g$, gauge pressure
Then $\ln k^{\#}(p_g) = \ln k^{\#}(\pi) + a_2 p_g/\pi_a + a_3 (p_g/\pi_a)^2$
Then $\{\ln [k^{\#}(p_g)/k^{\#}(\pi_a)]\}/p_g = a_2/\pi_a + a_3 p_g/\pi_a^2$
Then, $\mathrm{d} \ln k^{\#}/\partial p_g = a_2/\pi_a + 2a_3 p_g/\pi_a^2 = -\Delta^{\neq} V^{\infty}(p_g)/(RT)$
At ambient pressure p_g = zero; $\Delta^{\neq} V^{\infty}(\pi_a) = -a_2 RT/\pi_a$

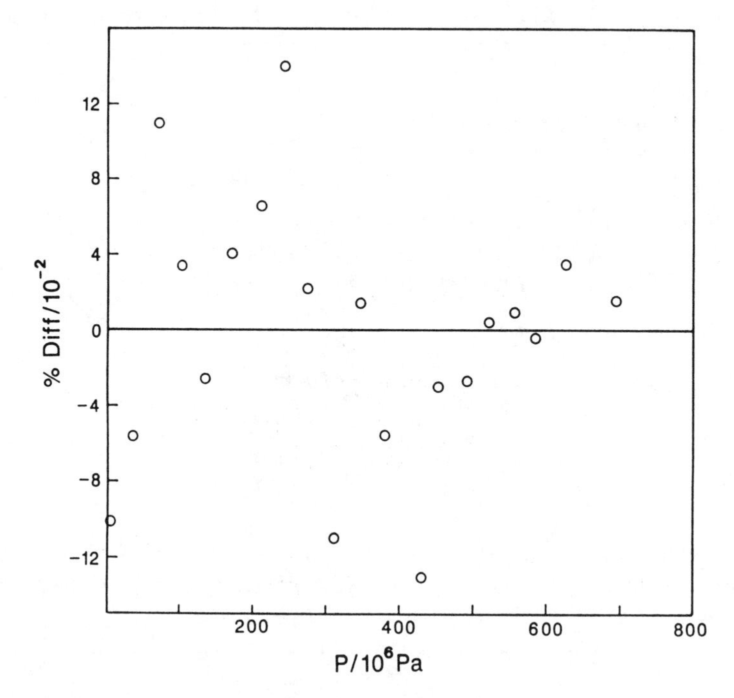

Fig. 4.6.1. Solvolysis of benzyl chloride(aq). Dependence on pressure of the percentage difference between $\ln k$(calc) and $\ln k$(obs) as a consequence of fitting to a quadratic in $p - \pi$.

in $p - \pi$. We illustrate (Table 4.6.7) the application by reference to the dependence on pressure of first order rate constants [5] for solvolysis of benzyl chloride(aq) (see also [10]). The limiting volume of activation is negative, the rate constant increasing with increase in pressure. A plot of the percentage difference between $\ln k^{\#}$(obs) and $\ln k^{\#}$(calc) against pressure shows no pattern (Fig. 4.6.1). Therefore we conclude that the equation describes satisfactorily the dependence of rate constant on pressure. The quadratic equation can be modified to the form shown in Table 4.6.8. This procedure is used when π is the ambient pressure and $p - \pi$ is the gauge pressure, p_g. The dependent variable, $\{\ln [k^{\#}(p_g)/k^{\#}(\pi)]/p_g\}$, is written as a linear function of p_g. Consequently a plot of $[\ln \{k^{\#}(p_g)/k^{\#}(\pi)\}/(p_g)$ against p_g yields a_2 as intercept and a_1 as the slope if the outcome of the analysis is a linear plot. This trend was observed

[12] for the second order rate constant for the reaction between ethyl iodide and triethylamine in methyl cyanide at 298 K.

References to section 4.6

[1] B. B. Owen; S. R. Brinkley. *Chem. Revs.*, 1941, **29**, 461; see equation (6).
[2] A. Disteche; S. Disteche. *J. Electrochem. Soc.*, 1967, **114**, 330.
[3] G. K. Ward; F. J. Millero. *Geochimica Cosmochimica Acta*, 1975, **39**, 1595.
[4] G. K. Ward; F. J. Millero. *J. Soln. Chem.*, 1974, **3**, 417.
[5] R. Lohmuller, D. D. Macdonald, M. Mackinnon and J. B. Hyne. *Canad. J. Chem.*, 1978, **567**, 1739.
[6] J. B. Hyne; H. S. Golinkin; W. G. Laidlaw. *J. Amer. Chem. Soc.*, 1966, **88**, 2104.
[7] H. S. Golinkin; W. G. Laidlaw; J. B. Hyne. *Canad. J. Chem.*, 1966, **44**, 2193.
[8] H. S. Golinkin; I. Lee; J. B. Hyne. *J. Amer. Chem. Soc.*, 1967, **89**, 1307.
[9] D. D. MacDonald; J. B. Hyne. *Canad. J. Chem.*, 1970, **48**, 2494.
[10] G. A. Lawrance, D. R. Stranks. *Inorg. Chem.*, 1978, **17**, 1804.
[11] P. O. I. Virtanen. *Suomen Kemistilehti*, 1967, **B40**, 179.
[12] Y. Kondo, H. Tojima, N. Tokura. *Bull. Chem. Soc. Japan*, 1967, **40**, 1408.

4.7 NORTH EQUATION

In the foregoing we used the measured dependence of $K(T)$ on pressure in order to calculate $\Delta_r V^\infty(T;p)$ and $\Delta_r \chi^\infty(T)$ for a given reaction. The aim is to use these derived quantities to say something about the chemistry of the substances involved in the reaction. In another approach we speculate about the chemistry and hence formulate an equation for the dependence of $K(T)$ on pressure. This is the procedure [1] adopted by North with respect to chemical equilibria in aqueous solutions. The starting hypothesis associates the dependence of $\Delta_r V^\infty$(aq) on pressure with the dependence on pressure of the molar volume of water (liquid 1) in the hydration sheaths of the solutes taking part in the chemical equilibrium. Thus $\Delta_r V^\infty(\text{aq}; T; p)$ is related (Table 4.7.1) to a parameter n which measures the change in the number of moles of water in the hydration sheath in the process reactants → products. The Tait equation (Chapter 1) is used to express the dependence of the molar volume of water on pressure. The outcome is a linear dependence of $\{[RT/(p - \pi)] \ln [K(p)/K(\pi)]\}$ on a function of a term which includes p, π and the Tait parameters. The slope yields an estimate of the hydration quantity n.

The compression quantity $\Delta_r \chi^\infty$ depends on pressure (Table 4.7.2). The form of the equation for $\Delta_r \chi^\infty$ is such that every subsequent derivative with respect to pressure is non-zero. An application of the North equation is summarized in Table 4.7.3. The analysis uses published Tait parameters [2] and density data [3] for water.

A similar analysis based on a modified Tait equation was discussed by Swaddle and coworkers [5] in the context of kinetic data. The experimental data concerns the first order rate constant for the aquation of pentaaminecobalt(III) complexes in aqueous solutions. The argument was advanced that on proceeding to the transition state h water molecules move from solvent water into the hydration sheath and in so doing

Table 4.7.1. North equation

At fixed temperature
Assumption: $\Delta_r V^\infty(\text{aq};p) = \Delta^{\neq} V^\infty(\text{aq};\pi) - n[V_1^*(1;p) - V_1^*(1;\pi)]$
From equation (1.7.4), $V_1^*(1;p) = V_1^*(1;\pi)\{1 - c\ln[(b+p)/(b+\pi)]\}$
or $V_1^*(1;p) - V_1^*(1;\pi) = -V_1^*(1;\pi)c\ln[(b+p)/(b+\pi)]$

$$\Delta_r V^\infty(\text{aq};p) = \Delta_r V^\infty(\text{aq};\pi) + nV_1^*(1;\pi)c\ln[(b+p)/(b+\pi)]$$

But $RT(\partial \ln K/\partial p) = -\Delta_r V^\infty(\text{aq};p)$. Then

$$RT\int_\pi^p \mathrm{d}\ln K = \int_\pi^p [-\Delta_r V^\infty(\text{aq};\pi) - nV_1^*(1;\pi)c\ln[(b+p)/(b+\pi)]]\,\mathrm{d}p$$

$$RT\int_\pi^p \mathrm{d}\ln K = \int_\pi^p [-\Delta_r V^\infty(\text{aq};\pi) - nV_1^*(1;\pi)c\{\ln[(b+p) - \ln(b+\pi)\}]\,\mathrm{d}p$$

But $\int \ln(b+p)\,\mathrm{d}p = (b+p)\ln(b+p) - p$

Hence

$$RT\int_\pi^p \mathrm{d}\ln K = [-\Delta_r V^\infty(\text{aq};\pi)p - nV_1^*(1;\pi)c\{(b+p)\ln(b+p) - p - p\ln(b+\pi)\}]_\pi^p$$

Then $RT\ln\{K(p)/K(\pi)\} = -\Delta_r V^\infty(\text{aq};\pi)(p-\pi) - nV_1^*(1;\pi)c\{(b+p)\ln(b+p) - (b+\pi)\ln(b+\pi) - p + \pi - p\ln(b+\pi) + \pi\ln(b+\pi)\}$

or $RT\ln\{K(p)/K(\pi)\} = -\Delta_r V^\infty(\text{aq};\pi)(p-\pi) - nV_1^*(1;\pi)c\{-(p-\pi) + (b+p)\ln[(b+p)/(b+\pi)]\}$

Hence, $[RT/(p-\pi)]\ln\{K(p)/K(\pi)\} = -\Delta_r V^\infty(\text{aq};\pi) + nV_1^*(1;\pi)c\{1 - [(b+p)/(p-\pi)]\ln[(b+p)/(b+\pi)]\}$

which has the form

$$[RT/(p-\pi)]\ln\{K(p)/K(\pi)\} = a_1 + a_2 V_1^*(1;\pi)c\{1 - [(b+p)/(p-\pi)] \times \ln[(b+p)/(b+\pi)]\}$$

where $a_1 = -\Delta_r V^\infty(\text{aq};\pi)$, and $a_2 = n$

Table 4.7.2. $\Delta_r \chi^\infty$ and the North equation

From Table 4.7.1 at fixed temperature

$$\Delta_r V^\infty(\text{aq};p) = \Delta_r V^\infty(\text{aq};\pi) + nV_1^*(1;\pi)c\ln[(b+p)/(b+\pi)]$$

$$\mathrm{d}\Delta_r V^\infty(\text{aq})/\mathrm{d}p = nV_1^*(1;\pi)c/(b+p)$$

Then $\Delta_r \chi^\infty = -nV_1^*(1;\pi)c/(b+p)^2$

Table 4.7.3. Ammonia(aq)

Equation: $NH_3(aq) + H_2O(aq) \rightleftharpoons NH_4^+(aq) + OH^-(aq)$
Data taken from [4].

$$1.01 \leqslant p/10^5\,Pa \leqslant 1.2159 \times 10^4$$

Temperature/K = 318.0; pressure π/Pa = 101 325.0; $K(aq;\pi) = 1.93 \times 10^{-5}$;
$\Delta_r V^\infty(aq;\pi) = -(26.31 \pm 0.07)\,cm^3\,mol^{-1}$; hydration parameter $n = 4.88 \pm 0.41$

Table 4.7.4. Swaddle Equation

At fixed temperature.
Chemical reaction: $MX_n(aq) + h\,H_2O \rightarrow [MX_n h\,H_2O(aq)]^{\neq}$
Modified Tait equation. At temperature T,

$$[V_1^*(1;\pi) - V_1^*(1;p)]/V_1^*(1;\pi) = c\ln[(b+p)/(b+\pi)]$$

Then $V_1^*(1;p) = V_1^*(1;\pi) - \{V_1^*(1;\pi)c[\ln(b+p) - \ln(b+\pi)]\}$

$$[\partial V_1^*(1)/\partial p] = -V_1^*(1;\pi)c/(b+p)$$

or $\Delta\chi^*(1) = V_1^*(1;\pi)c/(b+p)$
Assumption: $\Delta^{\neq}\chi^\infty = -hV_1^*(1;\pi)c/(b+p) = -\{\partial\Delta^{\neq}V^\infty(aq)/\partial p\}$

or $\int_\pi^p d\Delta^{\neq}V^\infty(aq) = hV_1^*(1;\pi)c\int_\pi^p\{1.0/(b+p)\}\,dp$

$$\Delta^{\neq}V^\infty(aq;p) = \Delta^{\neq}V^\infty(aq;\pi) + hV_1^*(1;\pi)c[\ln(b+p)]_\pi^p$$

$$\Delta^{\neq}V^\infty(aq;p) = \Delta^{\neq}V^\infty(aq;\pi) + hV_1^*(1;\pi)c\ln\{(b+p)/(b+\pi)\}$$

But $\int_\pi^p d\ln k^{\#}(aq) = -(1/RT)\cdot\int_\pi^p \Delta^{\neq}V^\infty(aq)\,dp$
Assumption; h = independent of pressure
Then $\ln[k^{\#}(aq;p)/k^{\#}(aq;\pi)] = -(1/RT)\Delta^{\neq}V^\infty(aq;\pi)(p-\pi)$
$-\{hV_1^*(1;\pi)c/(RT)\}\int_\pi^p[\ln(b+p) - \ln(b+\pi)]\,dp$

$$\int_\pi^p[\ln(b+p) - \ln(b+\pi)]\,dp = [(b+p)\ln(b+p) - (b+p) - p\ln(b+\pi)]_\pi^p$$

$$= (b+p)\ln(b+p) - (b+\pi)\ln(b+\pi) - b - p + b + \pi$$

$$-p\ln(b+\pi) + \pi\ln(b+\pi)$$

$$= (b+p)\ln[(b+p)/(b+\pi)] - (p-\pi)$$

Hence $\ln[k(aq;p)/k(aq;\pi)] = -(1/RT)\Delta^{\neq}V^\infty(aq;\pi)(p-\pi)$
$-\{hV_1^*(1;\pi)c/(RT)\}\{(b+p)\ln[(b+p)/(b+\pi)] - (p-\pi)\}$

where at 298.2 K, $V_1^*(1; H_2O; 101\,325\,Pa)/m^3\,mol^{-1} = 18.0 \times 10^{-6}$, $b = 3.06\cdot10^3$ bar and $c = 0.321$

NB: The final equation differs from that given by Swaddle because we have used reference pressure π.

Table 4.7.5. Aquation of $[Co(NH_3)_5NO_3]^{2+}$ (aq)

Temperature/K = 298.2 K. Reference pressure $\pi/\text{Pa} = 1.0 \times 10^5$
Data from [5]. Pressure range $1.0 \times 10^5 \leqslant p/\text{Pa} \leqslant 4.05 \times 10^8$
Rate constant: $k(\pi)/\text{s}^{-1} = 2.38 \times 10^{-5}$
$\Delta^{\neq} V^{\infty}(\text{aq}; \pi)/\text{cm}^3\,\text{mol}^{-1} = -6.17 \pm 0.24$ $\Delta h = 1.85 \pm 0.26$

lose their compressibility. The starting point is a modified Tait equation describing the dependence of molar volume of water on pressure (Table 4.7.4). An application is reviewed in Table 4.7.5.

References to section 4.7
[1] N. A. North. *J. Phys. Chem.*, 1973, **77**, 931.
[2] R. E. Gibson; O. H. Loeffler. *J. Amer. Chem. Soc.*, 1941, **63**, 898.
[3] R. A. Fine; F. J. Millero. *J. Chem. Phys.*, 1973, **59**, 5529.
[4] S. D. Hamann; W. Strauss. *Trans. Faraday Soc.*, 1955, **57**, 1684.
[5] W. E. Jones, L. R. Carey; T. W. Swaddle. *Canad. J. Chem.*, 1972, **50**, 2739.

4.8 BENSON AND BERSON EQUATION

In another development [1] the Tait equation (Chapter 1) for the compression of a liquid k_c is written in the form shown in Table 4.8.1. Assuming $b \gg \pi$, the reference pressure, the ratio of compressions at two pressures $k_c(p_1)/k_c(p_2)$ is given by a ratio

Table 4.8.1. Tait equation

Single substance; 1 mole of liquid 1_1.
At pressure p, $V_1^* = V_1^*[T; p]$. At reference pressure π, $V_1^* = V_1^*[T; \pi]$

$$k_c = \{V_1^*(\pi) - V_1^*(p)\}/V_1^*(\pi)$$

From Chapter 1, Table 1.7.4, (Tait Equation)

$$k_c = \{V_1^*(\pi) - V_1^*(p)\}/V_1^*(\pi) = c \ln\{(p + b]/[\pi + b]\}$$

Then at pressures p_1 and p_2;

$$k_c(p_1)/k_c(p_2) = \{\ln(p_1 + b) - \ln(\pi + b)\}/\{\ln(p_2 + b) - \ln(\pi + b)\}$$

Assumption: $k_c(p_1)/k_c(p_2) = 2$
Hence $(p_2 + b)^2/(\pi + b)^2 = (p_1 + b)/(\pi + b)$
or $(p_1 + b)(\pi + b) = (p_2 + b)^2$
Assumption: $b \gg \pi$. Then, $bp_1 + b^2 = p_2^2 + 2bp_2 + b^2$
or $p_2^2 = b[p_1 - 2p_2]$
Then, $b = p_2^2/(p_1 - 2p_2)$

Table 4.8.2. Benson–Berson equation

At fixed temperature.
Chemical reaction between r solutes in solution:

$$-RT[\partial \ln k^{\#}/\partial p]_T = V_{\neq}^{\infty}(\text{sln}; p) - \Sigma(j = 1; j = r)\, V_j^{\infty}(\text{sln}; p)$$

For the transition state using the Tait equation

$$V_{\neq}^{\infty}(\text{sln}; p) = V_{\neq}^{\infty}(\text{sln}; \pi) - V_{\neq}^{\infty}(\text{sln}; \pi) C \ln\{(p + b_{\neq})/(\pi + b_{\neq})\}$$

For reactant j using the Tait equation

$$V_j^{\infty}(\text{sln}; p) = V_j^{\infty}(\text{sln}; \pi) - V_j^{\infty}(\text{sln}; \pi) C \ln\{(p + b_j)/(\pi + b_j)\}$$

Hence, $-RT[\partial \ln k^{\#}/\partial p]_T = V_{\neq}^{\infty}(\text{sln}; \pi) - \Sigma(j = 1; j = r) V_j^{\infty}(\text{sln}; \pi)$
$- V_{\neq}^{\infty}(\text{sln}; \pi) C \ln\{(p + b_{\neq})/(\pi + b_{\neq})\}$
$+ \Sigma(j = 1; j = r) V_j^{\infty}(\text{sln}; \pi) C \ln\{(p + b_j)/(\pi + b_j)\}$
Assumption: b_j = constant for all r reactants
Also, $V_t^{\infty}(\text{sln}; p) = \Sigma(j = 1; j = r) V_j^{\infty}(\text{sln}; p)$
Hence, $-RT[\text{d} \ln k^{\#}/\partial p]_T = \Delta^{\neq} V^{\infty}(\text{sln}; \pi) - V_{\neq}^{\infty}(\text{sln}; \pi) C \ln\{(p + b_{\neq})/(\pi + b_{\neq})\}$
$+ V_t^{\infty}(\text{sln}; \pi) C \ln\{(p + b_j)/(\pi + b_j)\}$
when $\Delta^{\neq} V^{\infty}(\text{sln}; \pi) = V_{\neq}^{\infty}(\text{sln}; \pi) - V_t^{\infty}(\text{sln}; \pi)$

Table 4.8.3. Benson–Berson equation integrated form

At temperature T. From Table 4.8.2,

$$-RT\{\ln[k^{\#}(p)/k^{\#}(\pi)]\} = \int_{\pi}^{p} [\Delta^{\neq} V^{\infty}(\text{sln}; \pi) - V_{\neq}^{\infty}(\text{sln}; \pi) C \ln\{(p + b_{\neq})/(\pi + b_{\neq})\}$$

$$+ V_t^{\infty}(\text{sln}; \pi) C \ln\{(p + b_j)/(\pi + b_j)\}]\, \text{d}p$$

Then, $\ln[k^{\#}(p)/\ln k^{\#}(\pi)] = -[\Delta^{\neq} V^{\infty}(\text{sln}; \pi)(p - \pi)/(RT)] - \text{I}$
where,

$$I = V_t^{\infty}(\text{sln}; \pi) C \int_{\pi}^{p} \left[\ln\left[\frac{(p + b_j)}{(\pi + b_j)}\right] - \left[\frac{V_{\neq}^{\infty}(\text{sln}; \pi)}{V_t^{\infty}(\text{sln}; \pi)}\right] \ln\left[\frac{(p + b_{\neq})}{(\pi + b_{\neq})}\right]\right] \text{d}p$$

But $\int \ln(x + b)\, \text{d}x = (x + b)\ln(x + b) - (x + b)$
and $\text{d}[(x + b)\ln(x + b) - (x + b)] = 1 + \ln(x + b) - 1$

of two logarithmic terms in pressure. Assuming this ratio equals 2, then b is calculated from p_1 and p_2.

For simple organic liquids (e.g. hydrocarbons and alkyl halides), the quantity C is effectively independent of liquid and temperature, i.e. $C = 0.214 \pm 0.008$. The

Table 4.8.4. Benson–Berson equation—final form

From Table 4.8.3, at temperature T

$$RT\{\ln[k^{\#}(p)/k^{\#}(\pi)\} = -\Delta^{\neq}V^{\infty}(\text{sln};\pi)(p-\pi)$$

$$-V_{\text{t}}^{\infty}(\text{sln};\pi)C\,[(p+b_j)\ln(p+b_j)-(p+b_j)-p\ln(\pi+b_j)]_{\pi}^{p}$$

$$+[V_{\neq}^{\infty}(\text{sln};\pi)/V_{\text{t}}^{\infty}(\text{sln};\pi)]C\,[(p+b_{\neq})\ln(p+b_{\neq})-(p+b_{\neq})-p\ln(\pi+b_{\neq})]_{\pi}^{p}$$

If $\beta = (p+b)\ln(p+b)-(p+b)-p\ln(\pi+b)-(\pi+b)\ln(\pi+b)+(\pi+b)$
$+\pi\ln(\pi+b)$
$=(p+b)\ln(p+b)-(p-\pi)-(p-b)\ln(\pi+b)$

Hence, $\beta = (p+b)\ln\{(p+b)/(\pi+b)\}-(p-\pi)$

Then by definition, $\tau(b) = \{(p+b)/(p-\pi)(\ln(p+b)/(\pi+b)\}-1$

Hence*, $\{1/(p-\pi)\}\{\ln[k^{\#}(p)k^{\#}(\pi)]\} = -[\Delta^{\neq}V^{\infty}(\text{sln};\pi)C/RT]$
$+\{V_{\text{t}}^{\infty}(\text{sln};\pi)C/RT\}\phi(\tau)$

where $\phi(\tau) = \tau(b_j)-[V_{\neq}^{\infty}(\text{sln};\pi)/V_{\text{t}}^{\infty}(\text{sln};\pi)]\tau(b_{\neq})$

The final equation differs from that in [1] because we have used reference pressure π.
If $\pi = 0$, $\tau(b) = [(p+b)/p]\ln[(p+b)/b]-1$
$=(1+1/x)\ln(1+x)-1$ where $x = p/b$

Table 4.8.5. Benson–Berson equation—linearized form

Reference pressure $\pi = 0$ ($\simeq 101\,325\ \text{N m}^{-2}$)

$$\{1/p)\ln\{k(p)/k(\pi)\} = -\{\Delta^{\neq}V^{\infty}(\text{s ln};\pi)/RT][1-0.008C]$$
$$-\{0.2CV_{\text{t}}^{\infty}(\text{s ln};\pi)/RT\}\{[1/b_j^{0.523}]$$
$$-[1+(\Delta^{\neq}V^{\infty}(\text{s ln};\pi)/V_{\text{t}}^{\infty}(\text{s ln};\pi))][1/b_{\neq}^{0.523}]\}\,p^{0.523}$$

Table 4.8.6. Benson–Berson Equation

Reference pressure π = zero.

$$(1/p)\ln\{k^{\#}(p)/k^{\#}(\pi)\} = a_1 + a_2p^{0.523}$$

Then, $\ln k^{\#}(p) = \ln k^{\#}(\pi) + a_1p + a_2p^{1.523}$

$$(\partial\ln k^{\#}/\partial p)_T = a_1 + (1.523)a_2p^{0.523}$$

$$\Delta^{\neq}V^{\infty}(\text{sln};p) = -RT[a_1 + 1.523a_2p^{0.523}]$$

$$(\partial\Delta^{\neq}V^{\infty}(\text{sln})/\partial p)_T = -0.7965RTa_2/p^{0.47}$$

assumption is made that these C parameters for apolar liquids also apply to the limiting partial molar volumes of apolar transition and initial states (Table 4.8.2).

The analysis [1] predicts (Table 4.8.3) that a plot of $[1/(p-\pi)]\ln\{k(p)/k(\pi)\}$ (at temperature T) against $\phi(\tau)$ has intercept equal to $-\Delta^{\neq}V^{\infty}(\text{sln};p)/RT$ and slope equal to $-V_{\text{t}}^{\infty}(\text{sln};p)C/RT$. Unfortunately to put this analysis to the test we require the quantities $b_{\neq}$ and b_{i} together with the ratio $V_{\neq}^{\infty}(\text{sln};\pi)/V_{\text{t}}^{\infty}(\text{sln};\pi)$. Two procedures cope with this problem. The first uses an iterative technique in which estimates of $b_{\neq}$ and b_{i} together with the volume ratio are refined. The second procedure uses an approximation of the function $\tau(b)$ where the reference pressure is zero (Table 4.8.4). Under these circumstances $\tau(b)$ is written as $\tau(x)$ where, $\tau(x)=\sigma x^{n}+\text{q}$. A plot of $\tau(x)$ against $\ln(x)$ is close to linear, providing the empirical relationship $\tau(x)=-0.2x^{0.523}$. In these terms, the Benson–Berson equation is re-expressed as shown in Table 4.8.5. Consequently a plot of $(1/p)\ln\{k(p)/k(\pi)\}]$ against $p^{0.523}$ is constructed, the slope and intercept being defined in Table 4.8.5. The equation given in Table 4.8.6 is referred to as the Benson–Berson equation [1]. A least squares analysis is used to obtain estimates a_1 and a_2. This procedure was used to analyse the pressure dependence of rate constants for a Diels–Alder dimerization of liquid isoprene where the data show that $\Delta^{\neq}V^{*}(348\text{ K})=-38.4\text{ cm}^{3}\text{ mol}^{-1}$.

References to section 4.8

[1] S. W. Benson; J. A. Berson. *J. Amer. Chem. Soc.*, 1962, **84**, 152; 1964, **86**, 259.

4.9 OWEN AND BRINKLEY EQUATIONS

The Owen–Brinkley low pressure equation (Table 4.6.1) assumes that the limiting volume of reaction is a linear function of pressure; i.e. $\Delta_{\text{r}}\chi^{\infty}$ is constant. This assumption may be unacceptable. In this case a modified Tait equation (see Chapter 1) describes [1] the dependence of limiting partial molar volume of a solute $V_j^{\infty}(\text{sln};T;p)$ on pressure (Table 4.9.1). The suggestion was made that the equation for $\chi_j^{\infty}(p)$ in Table 4.9.1 is valid to higher pressures than the Tait equation. Parameter B (expressed in N m^{-2}) is characteristic of solvent and temperature. The outcome is an equation for the dependence on pressure at temperature T of $\ln K(p)$ about a reference pressure π in terms of three unknowns: $\ln K(\text{sln};\pi)$, $\Delta_{\text{r}}V^{\infty}(\text{sln};\pi)$, and $\Delta^{\neq}\chi^{\infty}(\text{sln};\pi)$. In their analysis, Owen and Brinkley quoted [1] the B-parameter for water at 298 K reported by Gibson [2]; i.e. 2996 bar.

Table 4.9.1. High pressure equation

Assumption (see Tait equation, Chapter 1). At temperature T.

$$\chi_j^{\infty}(\text{sln};p) = \chi_j^{\infty}(\text{sln};\pi)\{[B+\pi]/[B+p]\}^2$$

Integration* between limits π and p:

$$V_j^{\infty}(\text{sln};T;p) - V_j^{\infty}(\text{sln};T;\pi) = -\chi_j^{\infty}(\pi)\{(B+\pi)(p-\pi)/(B+p)\}$$

Integration check: $(B+\pi)(\mathrm{d}/\mathrm{d}p)[(p-\pi)(B+p)^{-1}]$

$$= (B+\pi)[\{1/(B+p)\} - \{(p-\pi)/(B+p)^2\}]$$

$$= (B+\pi)[(B+p)-(p-\pi)]/(B+p)^2 = (B+\pi)^2/(B+p)^2$$

Then,

$$\ln K(\text{sln};p) - \ln K(\text{sln};\pi) = -(1/RT)\int_{\pi}^{p}\{\Delta_r V^{\infty}(\text{sln};\pi) - \Delta_r\chi^{\infty}(\pi) \times [(B+\pi)(p-\pi)/(B+p)]\}\,\mathrm{d}p$$

Hence*,

$$\ln K(\text{sln};p) = \ln K(\text{sln};\pi) - \Delta_r V^{\infty}(\text{sln};\pi)(p-\pi)/RT + \{\Delta_r\chi^{\infty}(\pi)/RT\}\{[(B+\pi)(p-\pi)] - [(B+\pi)^2\ln[(B+p)/(B+\pi)]]\}$$

or

$$\ln K(\text{sln};p) = \ln K(\text{sln};\pi) - \Delta_r V^{\infty}(\text{sln};\pi)[(p-\pi)/RT] + \Delta_r\chi^{\infty}(\pi)\{[(B+\pi)(p-\pi)] - (B+\pi)^2\ln[(B+p)/(B+\pi)]\}/RT$$

Integration check: $(B+\pi)(\mathrm{d}/\mathrm{d}p)[(p-\pi) - (B+\pi)\ln\{(B+p)/(B+\pi)\}]$
$= (B+\pi)[1-(B+\pi)/(B+p)] = (B+\pi)[\{(B+p)-(B+\pi)\}/(B+p)]$
$= (B+\pi)(p-\pi)/(B+p)$

*Units: $[1] = [1] - [\mathrm{m^3\,mol^{-1}}][\mathrm{N\,m^{-2}}]/[\mathrm{J\,mol^{-1}\,K^{-1}}][\mathrm{K}]$
$+ [\mathrm{m^3\,mol^{-1}}\,(\mathrm{N\,m^{-2}})^{-1}][\mathrm{N\,m^{-2}}]^2/[\mathrm{J\,mol^{-1}\,K^{-1}}][\mathrm{K}]$

Footnote: We have modified the original treatment: here π is a reference pressure rather than 1 atm.

References to section 4.9

[1] B. B. Owen; S. R. Brinkley. *Chem. Rev.*, 1941, **29**, 461.
[2] R. E. Gibson. *Am. J. Sci.*, 1938, **35A**, 39.

4.10 LOWN, THIRSK AND WYNNE-JONES EQUATION

According to the analysis described in Table 4.9.1, the dependence, at constant temperature, of $\ln K(\text{sln};p)$ on pressure is characterized by $\ln K(\text{sln};\pi)$, $\Delta_r V^{\infty}(\text{sln};\pi)$ and $\Delta_r\chi^{\infty}$. The analysis is simplified if two of these parameters are related. For weak

Table 4.10.1. Lown, Thirsk and Wynne-Jones equation

At temperature T

From Table 4.6.1, $\ln K(\mathrm{sln};p) = \ln K(\mathrm{sln};\pi) - \Delta_r V^\infty(\mathrm{sln};\pi)[(p-\pi)/RT] + \Delta_r \chi^\infty[(p-\pi)^2/2RT]$

But $\Delta_r V^\infty(\mathrm{sln};\pi)/\Delta_r\chi^\infty$ = a constant, q (say)

$$\text{e.g. } \Delta_r\chi^\infty = q^{-1}\Delta_r V^\infty(\mathrm{sln};\pi)$$

Hence, $\ln K(\mathrm{sln};p) = \ln K(\mathrm{sln};\pi) - \Delta_r V^\infty(\mathrm{sln};\pi)\{[(p-\pi)/(RT)] - [(p-\pi)^2/(2qRT)]\}$

Table 4.10.2. Ethanoic acid(aq)

Reference [2]. Temperature/K = 298.2 Reference pressure/Pa = 1.0×10^5

Pressure range: $1.0 \times 10^5 \leqslant p/\mathrm{Pa} \leqslant 3.0 \times 10^8$

$\ln K(\mathrm{aq};\pi) = -10.984 \pm 0.024$, $\Delta_r V^\infty(\mathrm{aq};\pi)/\mathrm{cm^3\,mol^{-1}} = -(13.03 \pm 0.046)$, standard error = 3.36×10^{-2}

acids in aqueous solutions the ratio $[\Delta_r V^\infty(\mathrm{sln};T;p=\text{ambient})/\Delta_r\chi^\infty]$ is often a constant, 4.67×10^8 Pa [1]. In this case, the dependence of $\ln K(\mathrm{sln};T;p)$ on pressure is described by two parameters, $\ln K(\mathrm{sln};T;\pi)$ and $\Delta_r V^\infty(\mathrm{sln};T;\pi)$ where π is ambient pressure (Table 4.10.1). An application of this equation to the dependence [2] of $K(\mathrm{aq};T)$ on pressure for ethanoic acid(aq) is described in Table 4.10.2.

References to section 4.10

[1] D. A. Lown; H. R. Thirsk; Lord Wynne-Jones. *Trans. Faraday Soc.*, 1968, **66**, 2073.

[2] D. A. Lown; H. R. Thirsk; Lord Wynne-Jones. *Trans Faraday Soc.*, 1970, **66**, 51.

4.11 ELY'ANOV AND GONIKBERG EQUATION

The starting point [1,2] for this method is an extrathermodynamic assumption (Table 4.11.1) based on the Tait equation (Chapter 1) which is used to express the dependence on pressure at fixed temperature of the limiting partial molar volume of

Table 4.11.1. El'yanov and Gonikberg Equation

At fixed temperature T.

For the limiting partial molar volume of solute i in solution

$$V_i^\infty(p) = V_i^\infty(\pi) - V_i^\infty(\pi)c_i\ln[(b_i+p)/(b_i+\pi)]$$

Assumption: for all i solutes, $b_i \gg \pi$

Hence $V_i^\infty(p) = V_i^\infty(\pi) - V_i^\infty(\pi)c_i\ln[1+(p/b_i)]$

or $V_i^\infty(p) = V_i^\infty(\pi) - V_i^\infty(\pi)c_i\ln[1+\beta_i p]$ where $\beta_i = 1/b_i$

Then $\mathrm{d}V_i^\infty/\mathrm{d}p = -V_i^\infty(\pi)c_i\beta_i/(1+\beta_i p)$

Table 4.11.2. Volume of reaction (fixed temperature)

At pressure p, $\Delta_r V^\infty(p) = \Sigma\,(i = 1; i = n)\, v_i\, V_i^\infty(p)$
Then from Table 4.11.1,

$$\Delta_r V^\infty(p) = \Delta_r V^\infty(\pi) - \Sigma\,(i = 1; i = n) v_i V_i^\infty(\pi) c_i \ln[1 + \beta_i p]$$

$$\text{or } \Delta_r V^\infty(p) = \Delta_r V^\infty(\pi)\left[1 - \frac{\Sigma\,(i = 1; i = n) v_i V_i^\infty(\pi) c_i \ln[1 + \beta_i p]}{\Delta_r V^\infty(\pi)}\right]$$

With $\alpha = [\Sigma\,(i = 1; i = n) v_i V_i^\infty(\pi) c_i]/\Delta_r V^\infty(\pi)$
Assumption: $\beta_i = \beta$ for all solutes
Then, $\Delta_r V^\infty(p) = \Delta_r V^\infty(\pi)\{1 - \alpha \ln[1 + \beta p]\}$
Hence, $RT \ln K(p) = RT \ln K(\pi) - \Delta_r V^\infty(\pi)(p - \pi) + \alpha \Delta_r V^\infty(\pi) I$

$$\text{where } I = \int_\pi^p \ln(1 + \beta p)\,\mathrm{d}p = \left[\frac{(1 + \beta p)}{\beta}\ln(1 + \beta p) - \frac{(1 + \beta p)}{\beta}\right]_\pi^p$$

Check: $\mathrm{d}[\;]/\mathrm{d}p = (\beta/\beta)\ln(1 + \beta p) + [(1 + \beta p)/\beta][\beta/(1 + \beta p)] - (\beta/\beta)$
$= \ln(1 + \beta p) + 1 - 1$

$$\mathrm{I} = \int_\pi^p \ln(1 + \beta p)\,\mathrm{d}p = \left[\left[\frac{(1 + \beta p)}{\beta}\right][\ln(1 + \beta p) - 1\right]_\pi^p$$

$$= \left[\left[\frac{(1 + \beta p)}{\beta}\right][\ln(1 + \beta p) - 1]\right] - \left[\left[\frac{(1 + \beta \pi)}{\beta}\right][\ln(1 + \beta \pi) - 1]\right]$$

$$= \left[\frac{(1 + \beta p)}{\beta}\right][\ln(1 + \beta p)] - \left[\frac{(1 + \beta \pi)}{\beta}\right][\ln(1 + \beta \pi)] - (1 + \beta p)/\beta + (1 + \beta \pi)/\beta$$

$$= \left[\frac{(1 + \beta p)}{\beta}\right][\ln(1 + \beta p)] - \left[\frac{(1 + \beta \pi)}{\beta}\right][\ln(1 + \beta \pi)] - (p - \pi)$$

Assumptions: $p \gg \pi$ and $\beta\pi \ll 1$

$$\text{Then } I = \left[\frac{(1 + \beta p)}{\beta}\right]\ln(1 + \beta p) - p$$

Hence, $\ln[K(p)/K(\pi)] = -[\Delta_r V^\infty(\pi)/RT]p + \alpha \Delta_r V^\infty(\pi)\{(1 + \beta p)/\beta\}\ln(1 + \beta p)$
$- \alpha p \Delta_r V^\infty(\pi)$
or $\ln[K(p)/K(\pi)] = -[\Delta_r V^\infty(\pi)/RT][(1 + \alpha)p - (\alpha/\beta)(1 + \beta p)\ln(1 + \beta p)]$

Table 4.11.3. Dependence of ln K(sln) on pressure

From Table 4.11.2, $\ln K(p) = \ln K(\pi) - \Delta_r V^{\infty}(\pi)\phi(p)$
where $\phi(p) = [1/RT][(1 + \alpha)p - (\alpha/\beta)(1 + \beta p)\ln(1 + \beta p)]$

Table 4.11.4. El'yanov and Gonikberg's extrathermodynamic assumptions

For two related chemical reactions A and B, at fixed temperature
(I) Chemical equilibria $[\partial \ln K(\mathrm{A; sln})/\partial p] = \beta[\partial \ln K(\mathrm{B; sln})/\partial p]$
(II) Chemical kinetics $[\partial \ln k(\mathrm{A; sln})/\partial p] = \beta[\partial \ln k(\mathrm{B; sln})/\partial p]$

solute i, V_i^{∞}(sln). The assumption is made that the parameter b_i is much larger than the reference pressure π. The final equation (Table 4.11.2) expresses $V_i^{\infty}(\mathrm{sln};p)$ as a function of $V_i^{\infty}(\mathrm{sln};\pi)$ and $\ln(1 + \beta_i p)$ where $\beta_i = 1/b_i$. If β_i. If β_i is constant for all i solutes in a given solvent, then $\ln K(\mathrm{sln};p)$ is expressed as a function of $\ln K(\mathrm{sln};\pi)$, $\Delta_r V^{\infty}(\pi, \mathrm{sln})$ and a quantity α together with an integral function of $\ln(1 + \beta p)$. The outcome (Table 4.11.3) is an equation for $\ln K(\mathrm{sln};p)$ as a function of $\ln K(\mathrm{sln};\pi)$ and a product of $\Delta_r V^{\infty}(\mathrm{sln};\pi)$ and a function, $\phi(p)$. The latter is a function of pressure and two further parameters, α and β. El'yanov and Gonikberg found that for 23 chemical equilibria involving non-ionic substances, $\alpha = 0.1790$ and $\beta = 3.91 \times 10^{-3}\,\mathrm{bar}^{-1}$. For 56 Diels–Alder reactions, the dependence of $\ln k(p)$ on pressure was accounted for using $\alpha = 0.170$ and $\beta = 4.94 \times 10^{-3}\,\mathrm{bar}^{-1}$.

The form of the equation derived in Table 4.11.3 prompts the extrathermodynamic relationships described in Table 4.11.4 which assert that for two rate or equilibrium constants describing similar chemical reactions, the differentials with respect to intensive variable p are simply related [3] (Table 4.11.4). Granted therefore that one has determined $\phi(p)$ for a given class of reaction then one may use $K(\pi)$ and $\Delta_r V^{\infty}(\pi)$ for a new member of this class to predict $K(p)$ at pressure p [4].

References to section 4.11

[1] B. S. El'yanov; E. M. Gonikberg. *J. Chem. Soc. Faraday Trans.*, I, 1979, **75**, 172.
[2] B. S. El'yanov; E. M. Vasylvitskaya. *Rev. Phys. Chem. Japan*, 1980, **50**, 169.
[3] B. S. El'yanov; M. Gonikberg. *Russ. J. Phys. Chem.*, 1962, **36**, 313.
[4] B. S. El'yanov; M. G. Gonikberg. *Russ. J. Phys. Chem.*,1972, **46**, 856.

4.12 EL'YANOV AND HAMANN EQUATION

The basis of this approach is the extrathermodynamic assumption discussed above. This assumption draws attention to the fact that in order to describe the dependence of $K(p)$ on pressure we need to define the function $\phi(p)$. Then the task resolves into identifying a satisfactory form which describes that variable $\phi(p)$. One possibility [1] is explored in Table 4.12.1 where we start with an equation for the dependence of $\Delta_r V^{\infty}(p)$ on pressure. The outcome (Table 4.12.2) is an equation for $\ln K(p)$ as a

Table 4.12.1. El'yanov and Hamann Equation

At fixed temperature.
Assumption: $\Delta_r V^\infty(p) = \Delta_r V^\infty(\pi)\phi(p)$ where $\phi(p) = (b+\pi)^2/(b+p)^2$
Assumption: $b \gg \pi$ with $\beta = 1/b$. Then $\phi(p) = 1/(1+\beta p)^2$

Then $\phi(p) = \int_\pi^p [1/(1+\beta p)^2]\,\mathrm{d}p = [-1/\{\beta(1+\beta p)\}]_\pi^p$

Check d[ ]/d$p = (1/\beta)(-1)(-1)\beta/(1+\beta p)^2 = 1/(1+\beta p)^2$

Then $\phi(p) = [-1/\beta]\,[[1/(1+\beta p)] - [1/(1+\beta\pi)]]$

Then $\phi(p) = [-1/\beta]\left[\dfrac{1+\beta\pi-1-\beta p}{(1+\beta p)(1+\beta\pi)}\right]$

If $p \gg \pi$, and if $\beta\pi \ll 1$, $\phi(p) = p/(1+\beta p)$

Table 4.12.2. Dependence of $K(p)$ on pressure

From $\ln[K(p)/K(\pi)] = -\phi(p)\Delta_r V^\infty(\pi)/(RT)$
Then $\ln K(p) = \ln K(\pi) - \Delta_r V^\infty(\pi)p/[RT(1+\beta p)]$
where $\phi(p) = p/(1+\beta p)$ or $p/\phi(p) = 1 + (\beta p)$

Table 4.12.3. Dependence of $K(p)$ on pressure

From Table 4.12.1, at fixed temperature

$$\phi(p) = 1/(1+\beta p)^2 \text{ then } \mathrm{d}\phi/\mathrm{d}p = -2\beta/(1+\beta p)^3$$

Then $\Delta_r\chi^\infty = -\mathrm{d}\Delta_r V^\infty/\mathrm{d}p = -\Delta_r V^\infty(\pi)(\mathrm{d}\phi/\mathrm{d}p)$
Hence $\Delta_r\chi^\infty = 2\Delta_r V^\infty(\pi)\beta/(1+\beta p)^2$

Table 4.12.4. 4-Methoxybenzoic acid(aq) [2]

Temperature/K = 298.2 Pressure π/Pa = 1.0×10^5
Pressure $1.0 \leqslant p/\mathrm{bar} \leqslant 3 \times 10^3$

$$\ln K(\pi) = -10.391 \pm 0.003,\ \Delta_r V^\infty(\pi)/\mathrm{cm^3\,mol^{-1}} = 12.07 \pm 0.05,$$

$$\text{st error} = 4.526 \times 10^{-3}$$

function of pressure. For aqueous solutions at 298.2 K, $\beta = 9.2 \times 10^{-5}\,\mathrm{bar^{-1}}$. The form of the equation requires that $\Delta_r\chi^\infty$ is non-zero and depends on pressure (Table 4.12.3). An example of the analysis is given in Table 4.12.4 with reference to

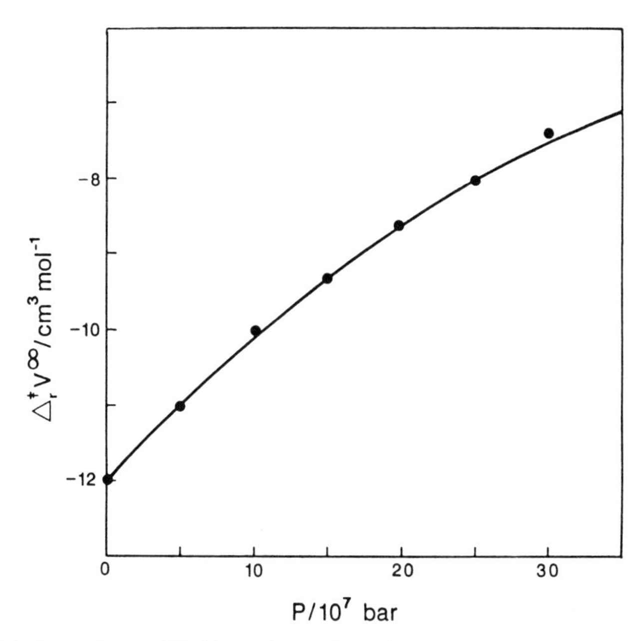

Fig. 4.12.1. Dependence of limiting volume of reaction, $\Delta_r V^{\infty}$(aq; 298.2 K) on pressure [2] for the acid dissociation of 4-methoxybenzoic acid(aq).

Table 4.12.5. El'yanov and Vasylvitskaya equation

At temperature T,

$$\Delta^{\neq} V(p) = \Delta^{\neq} V^{\infty}(\pi)[1 - a \ln\{1 + b(p - \pi)\}]$$

Then, $\ln k^{\#}(p) - \ln k^{\#}(\pi) = -[\Delta^{\neq} V^{\infty}(\pi)/RT](p - \pi) + [a\Delta^{\neq} V^{\infty}(\pi)/RT]I$

where $I = \int_{\pi}^{p} \ln[1 + b(p - \pi)]\, \mathrm{d}p$

$$= [\{[1 + b(p - \pi)]/b\} \ln[1 + b(p - \pi)] - (p - \pi)]$$

Check $\mathrm{d}[\ ]/\mathrm{d}p = (b/b)\ln[1 + b(p - \pi)] + \{[1 + b(p - \pi)]/b\}\{b/[1 + b(p - \pi)]\} - 1$
$= \ln[1 + b(p - \pi)]$

Then $\ln k^{\#}(p) = \ln k^{\#} - \{\Delta^{\neq} V^{\infty}(\pi)/RT\}(p - \pi)$
$+ \{a\Delta_r V^{\infty}(\pi)/RT\}\{\{[1 + b(p - \pi)]/b\} \ln[1 + b(p - \pi)] - (p - \pi)\}$

If $p \gg \pi$,

$$\ln k^{\#}(p) = \ln k^{\#}(\pi) - \{\Delta^{\neq} V^{\infty}(\pi)/RT\}p$$

$$+ \{a\Delta_r V^{\infty}(\pi)/RT\}\{\{[1 + bp]/b\} \ln[1 + bp] - p\}$$

or $\ln k^{\#}(p) = \ln k^{\#}(\pi) - \{\Delta^{\neq} V^{\infty}(\pi)/RT\}\{p(1 + a) - (a/b)\{[1 + bp]/b \ln[1 + bp]\}\}$

Table 4.13.1. Nakahara equation

$$\ln K(\mathrm{sln};p) = \ln K(\mathrm{sln};T;\pi) - [\Delta_r V^\infty(\mathrm{sln};T;\pi)/RT]\, b \ln\{1 + [(p-\pi)/b]\}$$

At $p = \pi$, $\ln K(\mathrm{sln};T;p) = \ln K(\mathrm{sln};T;\pi)$.

$$\Delta_r V^\infty(\mathrm{sln};T;p) = \Delta_r V^\infty(\mathrm{sln};T;\pi) b/[b + (p-\pi)]$$

the acid dissociation constant of 4-methoxybenzoic acid. The calculated dependence of limiting volume of reaction is summarized in Fig. 4.12.1.

The form of the function $\phi(p)$ lies at the heart of this approach. El'yanov and Vasylvitskayo [3] formulate the dependence [4] of $\Delta_r V^\infty(p)$ on pressure as shown in Table 4.12.5.

References to section 4.12

[1] B. S. El'yanov; S. D. Hamann. *Aust. J. Chem.*, 1975, **28**, 945.
[2] A. R. Fischer; B. R. Mann; J. Vaughan. *J. Chem. Soc.*, 1961, 1093.
[3] B. S. El'yanov; E. M. Vasylvitskaya. *Rev. Phys. Chem. Japan*, 1980, **50**, 169.
[4] M. V. Basilevsky; N. N. Weisberg; V. M. Zhulin. *J. Chem. Soc. Faraday Trans.*, *1*, 1985, **81**, 875.

4.13 NAKAHARA EQUATION

A series of arguments similar to that described in the previous section prompted a description of the dependence of $K(T;p)$ on pressure using equations (Table 4.13.1) based on the Owen–Brinkley equation for the dependence of the relative permittivity of a solvent ε_r on pressure.

References to section 4.13

[1] M. Nakahara, *Rev. Phys. Chem. Japan*, 1975, **66**, 57.

4.14 ASANO EQUATION

Often equations expressing the dependence of $\ln k^{\neq}$ on pressure require a pattern to derived parameter at high pressures which is unexpected based on the trends at low pressures, e.g. a change in sign of $\{\partial\Delta^{\neq} V^\infty(\mathrm{sln};p)/\partial p\}_T$. A key feature of the analyses concerns the estimate of $\Delta^{\neq} V^\infty(\mathrm{sln};T;p)$ at ambient pressure, the lowest pressure. With improvements in technology, so the upper pressure limit for kinetic data has increased. In other words the mean pressure over the experimental range has moved to higher pressures and away from ambient pressure. To a first approximation, a linear least squares analysis fits the dependence of $\ln k^{\neq}$ on pressure about the mean pressure. Again to a first approximation the standard error on a derived $\Delta^{\neq} V^\infty(\mathrm{sln};T;p)$ increases as one moves away from the mean temperature. This trend is particularly unwelcome if one aim of the experiment is to measure $\Delta^{\neq} V^\infty(\mathrm{sln};T)$ at ambient pressure. Asano sought to formulate equations which minimized these and related problems. Several criteria were demanded of equations relating rate constant

Table 4.14.1. Asano criteria

(i) Equation must reproduce measured dependence of rate constant $k^{\#}$ on pressure.
(ii) limit $(p \rightarrow \pi)$ $\mathrm{d}\ln k^{\#}/\mathrm{d}p$ = independent of pressure.
(iii) limit $(p \rightarrow \infty)$ $\mathrm{d}\ln k^{\#}/\mathrm{d}p$ = constant.
(iv) Equation must be simple.

Table 4.14.2. Asano equation (A)

At fixed temperature T; reference pressure = π

$$\ln(k^{\#}(p)\} = \ln k^{\#}(\pi)) + a(p-\pi) + b(p-\pi)/[1+c(p-\pi)]$$

$$[\partial \ln k^{\#}/\partial p] = a + \{b/[1+c(p-\pi)]\} - \{cb(p-\pi)][1+c(p-\pi)]^2\}$$

$$= a + \{b/[1+c(p-\pi)]^2\}$$

Hence $\Delta^{\neq} V^{\infty}(\mathrm{sln}) = -RT[a + \{b/[1+c(p-\pi)]^2\}]$

$$\text{limit}(p \rightarrow \pi)\, \mathrm{d}\ln k^{\#}/\mathrm{d}p = a + b \text{ or } \Delta^{\neq} V^{\infty}(\mathrm{sln}; \pi) = -RT(a+b)$$

Then $a = -\{[\Delta^{\neq} V^{\infty}(\mathrm{sln}; \pi)/RT] + b\}$ Also limit $(p \rightarrow \infty)\, \mathrm{d}\ln k^{\#}/\mathrm{d}p = a$

Table 4.14.3. Asano equation (B)

Conditions: see Table 4.14.1
At fixed temperature,

$$\ln\{k^{\#}(p)\} = \ln k^{\#}(\pi) + a(p-\pi) + b\ln[1+c(p-\pi)]$$

$$\{\partial \ln k^{\#}/\partial p\} = a + \{bc/[1+c(p-\pi)]\}$$

$$\text{limit}(p \rightarrow \infty)\, \mathrm{d}\ln k^{\#}/\mathrm{d}p = a;\ \text{a constant}$$

$$\text{limit}(p \rightarrow \pi)\, \mathrm{d}\ln k^{\#}/\mathrm{d}p = a + (bc);\ \text{a constant}$$

Therefore $a + bc = -\Delta^{\neq} V^{\infty}(\mathrm{sln}; \pi)/RT$
or $a = -[\Delta^{\neq} V^{\infty}(\mathrm{sln}; \pi)/RT] - (bc)$
Hence, $\ln\{k^{\#}(p)\} = \ln\{k^{\#}(\pi)\} - \{\Delta^{\neq} V^{\infty}(\mathrm{sln}; \pi)/RT + bc\}(p-\pi)$
$+ b\ln[1+c(p-\pi)]$
Also $\Delta^{\neq} V^{\infty}(\mathrm{sln}; p) = -RT[a + \{bc/[1+c(p-\pi)]\}]$
and $\Delta^{\neq} \chi^{\infty}(\mathrm{sln}; p) = -RTbc^2/[1+c(p-\pi)]^2$

$k^{\#}(p)$ with pressure (Table 4.14.1). The procedures adopted here involve starting out with a possible equation and testing against the criteria in Table 4.14.1, particularly criteria (ii) and (iii). Possible equations are explored in Tables 4.14.2 and 4.14.3. These equations describe satisfactorily the dependence on pressure of rate constants for several Diels–Alder reactions. The parameter 'a', which corresponds to the limit

Table 4.15.1. Marshall and Mesmer Equation

At fixed temperature T; $\ln K(p) = \ln K(\pi) + n \ln [\rho_1^*(p)/\rho_1^*(\pi)]$
where ρ_1^* = density of the solvent
Then $\mathrm{d} \ln K/\mathrm{d}p = n \,\mathrm{d} \ln \rho_1^*/\mathrm{d}p$, or $\Delta_r V^\infty(p) = -RTn\kappa_1^*$

$(p \to \infty)\, \mathrm{d} \ln k^{\#}/\mathrm{d}p$, was understood to represent the intrinsic volume of activation, namely the volume of activation minus the contribution made by the solvent. The major disadvantage of the analysis is that the fitting procedures require non-linear curve-fitting techniques in order to estimate parameters a, b and c (see for example [2]).

References to section 4.14

[1] T. Asano; T. Okada. *J. Phys. Chem.*, 1984, **88**, 238.
[2] W. E. Wentworth. *J. Chem. Educ.*, 1965, **42**, 96.

4.15 MARSHALL AND MESMER EQUATION

Part of the interest in probing the dependence of equilibrium (and rate) constants on pressure emerges from attempts to understand the role of the solvent in these processes. The argument proceeds from the dependence on p of $\ln K(\mathrm{s\,ln}; p)$ through the calculated quantities $\Delta_r V^\infty$ and $\Delta_r \chi^\infty$ towards modelling solvent–solute interactions. In an elegant development Marshall and coworkers [1, 2] explored directly the dependence of $\ln K(T)$ on the density (or, molar volume) of the solvent (Table 4.15.1)

References to section 4.15.1

[1] W. L. Marshall; R. E. Mesmer. *J. Soln. Chem.*, 1981, **10**, 121.
[2] W. L. Marshall; R. E. Mesmer. *J. Soln. Chem.*, 1984, **13**, 383.

5

Dependence of equilibrium and rate constants on temperature—linear equations

5.1 INTRODUCTION

The partial differential of $\ln K(T;p)$ with respect to temperature is related to the limiting enthalpy of reaction, $\Delta_r H^\infty(\text{sln}; T; p)$ by the van't Hoff equation. The partial differential of $\Delta_r H^\infty(\text{sln}; p)$ with respect to temperature is related to the limiting isobaric heat capacity of reaction, $\Delta_r C_p^\infty(\text{sln})$. For most purposes ambient pressure is sufficiently close to p^0 that $K(\text{sln}; T; \text{ambient } p)$ equals $K^0(\text{sln}; T)$.

We assume that at pressure p and temperature T there exists a true equilibrium constant $K(\text{sln}; \text{true}; T; p)$ (Table 3.1.4). Experimental data comprise estimates $K(\text{sln}; T; p)$ at different temperatures, the latter being generally assumed without error. We assert there is a pattern to the dependence of $K(\text{sln}; \text{true}; T; p)$ on T; i.e. for a given system $K(\text{sln}; \text{true}; T; p)$ at T_1 is related to $K(\text{sln}; \text{true}; T; p)$ at T_2. The task is to examine the dependence of estimated $K(\text{sln}; T; p)$ on T and hence discover a quantitative relationship between $K(\text{sln}; \text{true}; T; p)$ and T.

In practice $K(\text{sln}; T; p)$ is an estimate at pressure p and at a discrete temperature within a limited temperature range. Limits on this range are set by human patience and available resources. Consequently, dependences are sampled at intervals of 5 K or larger rather than at intervals of 0.1 K. For aqueous solutions at ambient pressure, the range $273.15 < T < 373.15$ K defines the window within which we may estimate the dependence of $K(\text{sln}; T; p)$ on T.

In surveying [1] procedures for analysing the dependence of $K(\text{sln}; T; p)$ on temperature we will find that many of the equations described in this chapter lead to a linear least squares analysis of experimental data. It is always tempting to push the data to the limits and, carried along with enthusiasm, to calculate a range of temperature derivatives including $d\Delta_r C_p^\infty/dT$, $d^2\Delta_r C_p^\infty/dT^2$ .... A proper statistical analysis counters this enthusiasm [2–4]. Some initial estimate should be made of the precision to which $K(\text{sln}; T; p)$ has been measured. Errors in $K(\text{sln}; T; p)$ propagate through derivatives with respect to temperature, affecting the reliability of derived parameters. Rarely are data sufficiently precise to warrant attaching significance [5] to $\Delta_r H^\infty(\text{sln}; T; p)$ smaller then $80\ \text{J mol}^{-1}$ and to $\Delta_r C_p^\infty(\text{sln}; T; p)$ smaller than

Table 5.1.1. Rate constants and temperature

From Table 3.2.4, $[\partial \ln (k^{\#}\text{K}/T)/\partial T]_p = \Delta^{\neq} H^{\infty}(\text{sln}; T; p)/RT^2$

$$\Delta^{\neq} C_{\text{p}}^{\infty}(\text{sln}; T; p) = [\partial \Delta^{\neq} H^{\infty}(\text{sln})/\partial T]_p$$

$$\Delta^{\neq} H^{\infty}(\text{sln}; T; p^0) = \Delta^{\neq} H^0(\text{sln}; T),$$

and $\Delta^{\neq} C_{\text{p}}^{\infty}(\text{sln}; T; p^0) = \Delta^{\neq} C_{\text{p}}^0(\text{sln}; T)$

$10\,\text{J}\,\text{K}^{-1}\,\text{mol}^{-1}$. Limiting enthalpies and isobaric heat capacities of reaction can also be independently estimated from calorimetric data. Therefore a cross-check on the analysis of the dependence of equilibrium constants on temperature is available. There are advantages in characterizing the equilibrium composition using the molality scale. If the equilibrium composition is characterized using concentrations rather than molalities, the dependence of solvent density on temperature must be incorporated into the analysis [6–8].

We also consider the task of analysing the dependence on temperature of rate constants for chemical reaction [9–13]. The starting point is the equation (Table 5.1.1) for $\text{d}\ln(k^{\#}\text{K}/T)/\text{d}T$ in terms of the limiting molar enthalpy of activation, $\Delta^{\neq} H^{\infty}(\text{sln}; T; p)$, at temperature T. (By defining a kinetic parameter $k^{\#}$ we avoid trailing units which might cause difficulties (see section 3.2).) The aim is to use the measured dependence of $k^{\#}$ on temperature in an attempt to formulate the dependence of the true $k^{\#}$ on temperature. At the outset we have no information concerning what this dependence might be. Having found a pattern to the dependence of $k^{\#}$ on T, we examine the dependence using transition state theory. In these terms the dependent variable is written as $\ln(k^{\#}\text{K}/T)$ rather than $\ln k^{\#}$. We assume the rate constant describes a single discrete process. In other words just one chemical process (or stage) is involved in the reaction reactants → products. Often we explore the dependence of $\ln(k^{\#}\text{K}/T)$ on T in order to calculate rate constants at selected temperatures. If this calculation involves interpolation, there are few problems. Where this calculation involves extrapolation outside the measured range, we must take note of the standard errors on the fitting parameters. The physical properties of a system define a window within which the rate constant can be measured. The window for kinetic studies is often smaller than that for equilibrium studies. The reaction may be too fast or too slow to be measured at extremes of the temperature range.

A contentious issue concerns isobaric heat capacities of activation, $\Delta^{\neq} C_{\text{p}}^{\infty}(\text{sln}; T; p)$. Kinetic data can be almost routinely [14, 15] obtained with sufficient precision to identify this quantity, namely the dependence of $\Delta^{\neq} H^{\infty}(\text{sln}; T; p)$ on temperature. Good data at six temperatures spanning a range of 50 K is often sufficient [14] to characterize a magnitude for $\Delta^{\neq} C_{\text{p}}^{\infty}(\text{sln}; T; p) \geqslant 40\,\text{J}\,\text{K}^{-1}\,\text{mol}^{-1}$. However, criteria for identifying dependences of $\Delta^{\neq} C_{\text{p}}^{\infty}(\text{sln}; T; p)$ on temperature must be rigorous from a statistical standpoint. Standard errors on derived activation parameters are functions of both the activation parameter and the temperature; these two dependences conspire to produce large standard errors on $\text{d}\Delta^{\neq} C_{\text{p}}^{\infty}/\text{d}T$.

References to section 5.1

[1] M. J. Blandamer; J. Burgess; R. E. Robertson; J. M. W. Scott. *Chem. Rev.*, 1982, **82**, 259.
[2] D. L. Albritton; A. L. Schmeltekopf; R. N. Zare. *Modern Spectroscopy, Modern Research II* (ed. K. N. Rao), Academic Press, New York, 1976.
[3] C. J. Brookes; I. G. Betteley; S. M. Loxston. *Fundamentals of Mathematics and Statistics*, Wiley, New York, 1979.
[4] B. A. Timimi. *Electrochim. Acta*, 1974, **19**, 149.
[5] E. J. King. *Acid–Base Equilibria*, Academic Press, New York, 1965.
[6] E. A. Guggenheim. *Trans. Faraday Soc.*, 1937, **33**, 607.
[7] L. G. Hepler. *Thermochimica Acta*, 1981, **50**, 69.
[8] E. K. Euranto; J. J. Kankare; N. J. Cleve. *J. Chem. Eng. Data*, 1969, **14**, 455.
[9] K. J. Laidler. *J. Chem. Educ.*, 1984, **61**, 494.
[10] S. R. Logan. *J. Chem. Educ.*, 1982, **59**, 279.
[11] H. Maskill. *Educ. in Chem.*, 1985, 154.
[12] J. R. Hulett. *Quart. Rev.*, 1964, **18**, 227.
[13] B. Perlmutter-Hayman. *Prog. Inorg. Chem.*, 1976, **20**, 229.
[14] R. E. Robertson. *Suomen Kemistilehti*, 1960, **33A**, 63. *Progr. Phys. Org. Chem.*, 1967, **4**, 213.
[15] G. Kohnstam. *Adv. Phys. Org. Chem.*, 1967, **5**, 121.

5.2 REFERENCE TEMPERATURES

The equilibrium acid dissociation constants at ambient pressure of many weak organic acids in aqueous solution [1–5] show extrema when plotted against temperature. Harned and Embree formulated [1, 2] the dependence of $K(\text{aq})$ on temperature (expressed in Celsius) in terms of $K(\theta)$ and $(T-\theta)$ where θ is the temperature (Celsius) at the extremum (Table 5.2.1). The a-parameter (Table 5.2.1) is assumed constant for aqueous solutions, independent of the acid. The equation for $\ln K(\text{aq}; T)$

Table 5.2.1. Harned and Embree general equation for weak acids in aqueous solution; chemical equilibria

At fixed pressure, p. If $\ln K$ has an extremum, $\ln K(\max)$, at θ Celsius, then in general terms,

$$\ln K = \ln K(\max) + f[(T-\theta)/\text{Celsius}]$$

Equation: $\ln K - \ln K(\max) = -a[(T-\theta)/\text{Celsius}]^2$

where $a = -2.17 \times 10^{-5}$

Analysis: $\ln K = \ln K(\max) - a[(T-\theta)/\text{Celsius}]^2$

or $\ln K = \ln K(\max) - a[T^2 - 2T\theta + \theta^2]/(\text{Celsius})^2$

Then, $\{\ln K + [a\theta^2/(\text{Celsius})^2]\} = 2a\theta T/(\text{Celsius})^2 + \ln K(\max) - aT^2/(\text{Celsius})^2$

Table 5.2.2. van't Hoff equation and reference temperature

From Table 3.1.5,

$$\int_{\theta}^{T} \mathrm{d}\ln K(\mathrm{sln}) = \int_{\theta}^{T} \{\Delta_{\mathrm{r}} H^{\infty}(\mathrm{sln})/RT^2\}\mathrm{d}T \qquad \text{(i)}$$

Also

$$\int_{\theta}^{T} \mathrm{d}\Delta_{\mathrm{r}} H^{\infty}(\mathrm{sln}) = \int_{\theta}^{T} \Delta_{\mathrm{r}} C_{\mathrm{p}}^{\infty}(\mathrm{sln})\mathrm{d}T \qquad \text{(ii)}$$

and

$$\int_{\theta}^{T} \mathrm{d}\Delta_{\mathrm{r}} S^{0}(\mathrm{sln}) = \int_{\theta}^{T} \Delta_{\mathrm{r}} C_{\mathrm{p}}^{\infty}(\mathrm{sln})\mathrm{d}\ln T \qquad \text{(iii)}$$

is rearranged such that a plot of $\ln K(\mathrm{aq}; T) + a(\theta/\mathrm{Celsius})^2$ against $T/\mathrm{Celsius}$ has slope equal to $2a\theta/(\mathrm{Celsius})$ and an intercept (at zero Celsius) equal to $\ln K(\max) - a(\theta/\mathrm{Celsius})^2$. Because a is known, $K(\max)$ can be calculated. This method is important in view of the suggestion concerning a reference temperature θ, but this temperature is not necessarily linked to $K(\max)$. In the following discussion, temperature θ is simply a reference temperature whereby $\ln K(\mathrm{sln}; T; p)$ at temperature T is compared with $\ln K(\mathrm{sln}; \theta; p)$ at temperature θ (at common pressure p).

The common basis of most treatments is the van't Hoff equation, which we write in integrated form between the limits temperature T and reference temperature θ (Table 5.2.2). The integrals on the right-hand side of equations (i)–(iii) of the latter table cannot be evaluated because we do not have explicit equations for these dependences.

A key problem concerns the heat capacities, $\Delta_{\mathrm{r}} C_{\mathrm{p}}^{\infty}(\mathrm{sln}; T; p)$ and $\Delta^{\neq} C_{\mathrm{p}}^{\infty}(\mathrm{sln}; T; p)$. If we could write down an equation describing the dependence of these quantities on temperature [6], the task of analysing the dependence of either $K(T)$ or $k^{\#}(T)$ on temperature would be straightforward. If $\Delta_{\mathrm{r}} C_{\mathrm{p}}^{\infty}$ is in fact zero, the integrations (Table 5.2.2) are trivial (section 5.3).

References to section 5.2

[1] H. S. Harned; N. D. Embree, *J. Amer. Chem. Soc.*, 1934, **56**, 1042.
[2] H. S. Harned; N. D. Embree. *J. Amer. Chem. Soc.*, 1934, **56**, 1050.
[3] H. S. Harned; G. L. Kazanjian. *J. Amer. Chem. Soc.*, 1936, **58**, 1912.
[4] H. S. Harned; R. W. Ehlers. *J. Amer. Chem. Soc.*, 1932, **54**, 1350.
[5] H. S. Harned; R. W. Ehlers. *J. Amer. Chem. Soc.*, 1933, **55**, 2379.
[6] I. Horsak; I. Slama. *Chemical Papers*, 1982, **36**, 745.

5.3 ELEMENTARY TREATMENT

Equilibrium composition

An obvious starting point assumes that $\Delta_{\mathrm{r}} C_{\mathrm{p}}^{\infty}(\mathrm{sln})$ is zero at all temperatures. In other words $\Delta_{\mathrm{r}} H^{\infty}(\mathrm{sln})$ is constant, independent of temperature over the range θ to T.

Table 5.3.1. Elementary treatment

$\Delta_r H^\infty$ is independent of temperature from Table 5.2.2, at fixed pressure p

$$\int_\infty^T \mathrm{d}\ln K(\mathrm{sln}) = \{\Delta_r H^\infty(\mathrm{sln})/R\}\int_\theta^T (1/T^2)\,\mathrm{d}T$$

Hence, $\ln K(\mathrm{sln}; T) = \ln K(\mathrm{sln}; \theta) - \{\Delta_r H^\infty(\mathrm{sln})/R\}[(1/T) - (1/\theta)]$

Table 5.3.2. Linear dependence of ln *K* on reciprocal temperature

From Table 5.3.1, assuming pressure $p \simeq p^0$.

$$\ln K(T) = \ln K(\theta) - \{\Delta_r H^\infty/R\}[(1/T) - (1/\theta)]$$

At $T/\mathrm{K} = 1.0$, $\ln K(T/\mathrm{K} = 1.0) = \ln K(\theta) + \{\Delta_r H^\infty/R\}[(1/\theta) - (1/\mathrm{K})]$
Then $\ln K(T) = \{\ln K(T/\mathrm{K} = 1.0) + \Delta_r H^\infty/R\mathrm{K}\} - [\Delta_r H^\infty/RT]$
But at $T/\mathrm{K} = 1.0$

$$\Delta_r H^\infty(\mathrm{sln}) - \mathrm{K}\Delta_r S^0(\mathrm{sln}) = -R\ln K(T/\mathrm{K} = 1)$$

or $\ln K(T/\mathrm{K} = 1) + \{\Delta_r H^\infty/R\mathrm{K}\} = \Delta_r S^0(\mathrm{sln})/\mathrm{R}$
Then, $\ln K(T) = a_1 + a_2\mathrm{K}/T$
where $a_1 = \Delta_r S^0(\mathrm{sln})/R$ and $a_2 = -\Delta_r H^\infty/R\mathrm{K}$

Table 5.3.3. Linear dependence of ln $k^\#$ on T^{-1}

At pressure p, $\ln k^\# = a_1 + a_2(\mathrm{K}/T)$; $\mathrm{d}\ln k^\#/\mathrm{d}T = -a_2\mathrm{K}/T^2$
But $\mathrm{d}\ln(k^\#\mathrm{K}/T)/\mathrm{d}T = \mathrm{d}\ln k^\#/\mathrm{d}T - (1/T) = \Delta^\# H^\infty/RT^2$
Therefore, $\Delta^\# H^\infty = -R(a_2\mathrm{K} + T)$; and, $\Delta^\# C_\mathrm{p}^\infty = -R$

Further the entropy of reaction $\Delta_r S(\mathrm{sln}; m_j = 1; \gamma_j = 1)$ is also independent of temperature. (Here we again use the symbol $\Delta_r S(\mathrm{sln}; m_j = 1; \gamma_j = 1)$ to refer to ideal solutions for each i solute.) The usual form of the equation derived in Table 5.3.1 is recovered if we assume that the equation is valid down to ! K (Table 5.3.2). Hence $\ln K(T)$ is a linear function of T^{-1}. A linear least squares analysis yields estimates of a_1 and a_2 which yield entropies and enthalpies of reaction respectively.

Kinetic data

The related analysis for kinetic data involves fitting the dependence of $\ln k^\#$ on T^{-1} to yield estimates [1] of a_1 and a_2 defined in Table 5.3.3. Although this dependence requires a non-zero $\Delta^\neq C_\mathrm{p}^\infty$ $(= -R)$, the magnitude of this heat capacity of activation is not experimentally meaningful relative to the precision to which rate constants can be measured.

References to section 5.3

[1] C. Rozycki. *J. Thermal Analysis*, 1984, **29**, 959.

5.4 VALENTINER EQUATION

Equilibrium composition

The restriction on the dependence of $\ln K(T)$ on T imposed by the equation in Table 5.3.1 is relaxed by adding further terms which are different functions of temperature. Many equations have the following general form:

$$\ln K(T) = a_1 + a_2(\mathrm{K}/T) + \Sigma(j = 3; j = \mathrm{i}) a_j f_j(T/\mathrm{K}) \tag{5.4.1}$$

The equation which adds one term linear in $\ln(T/\mathrm{K})$ is identified by a number of names, e.g. the Valentiner equation [1] and the Everett–Wynne-Jones equation [2]. Application of this equation to systems of chemical interest was advocated by Everett [2–6] and by Moelwyn-Hughes [7, 8]. Most applications have concerned chemical equilibria in aqueous solutions [9–12].

Here we develop the argument, starting with the assumption that the limiting isobaric heat capacity of reaction $\Delta_r C_p^\infty(T)$ is independent of temperature (Table 5.4.1). Then two integrations between limits T and reference temperature θ yields an equation for $\ln K(T)$ in terms of the limiting enthalpy of reaction at temperature θ, together with the limiting isobaric heat capacity of reaction $\Delta_r C_p^\infty$. The resulting equation has two temperature functions: $(1/\theta) - (1/T)$ and $\ln(T/\theta) + (\theta/T) - 1$. A linear least squares analysis of the dependence on T of $\ln K(T)$ yields estimates of $\Delta_r H^\infty(\theta)$ and $\Delta_r C_p^\infty$.

At an extremum (at T_{ex}) in the dependence of $\ln K(T)$ on T, the enthalpy of reaction is zero. Hence the calculated enthalpies and isobaric heat capacities can be used to

---

Table 5.4.1. Chemical equilibria

At pressure p,

$$\int_\theta^T \mathrm{d}\Delta_r H^\infty(\mathrm{sln}) = \int_\theta^T \Delta_r C_p^\infty(\mathrm{sln})\mathrm{d}T$$

Assumption: $\Delta_r C_p^\infty = \text{constant}$

$$\Delta_r H^\infty(\mathrm{sln}; T) = \Delta_r H^\infty(\mathrm{sln}; \theta) + \Delta_r C_p^\infty(T - \theta)$$

Then (Table 5.3.1)

$$\int_\theta^T \mathrm{d}\ln K = \int_\theta^T [[\Delta_r H^\infty(\theta)/RT^2] + [\Delta_r C_p^\infty/R][(1/T) - (\theta/T^2)]]\mathrm{d}T$$

Hence, $\ln K(T) = \ln K(\theta) + [[\Delta_r H^\infty(\theta)/R](-1/T) + [\Delta_r C_p^\infty/R][\ln T + (\theta/T)]]_\theta^T$

Therefore, $\ln K(T) = \ln K(\theta) + [\Delta_r H^\infty(\theta)/R][(1/\theta) - (1/T)]$
$\qquad + [\Delta_r C_p^\infty/R][\ln(T/\theta) + (\theta/T) - 1]$

Table 5.4.2. Extrema in ln K as a function of T

From Table 5.4.1, if at T_{ex}, $\Delta_r H^\infty(T) = 0.0$, then $-\Delta_r H^\infty(\theta) = \Delta_r C_p^\infty(T_{ex} - \theta)$
Then $T_{ex} = \{-\Delta_r H^\infty(\theta) + \theta\Delta_r C_p^\infty\}/\Delta_r C_p^\infty$
From Table 5.4.3, if $\mathrm{d}\ln K/\mathrm{d}T = 0$, $\ln K$ is an extremum
where $0 = -a_2(\mathrm{K}/T_{ex}^2) + (a_3/T_{ex})$ or $T_{ex} = a_2(\mathrm{K}/a_3)$,

estimate temperature T_{ex} (Table 5.4.2). The equation described in Table 5.4.1 does not have the form usually associated with the name Valentiner. However, the classic form of the equation is recovered (Table 5.4.3) if we combine equations for $\ln K(T)$ at temperature T and at $T/\mathrm{K} = 1$. Then a linear least squares analysis of the dependence of $\ln K(T)$ on T yields estimates of a_1, a_2 and a_3 together with their standard errors. This is the route often taken. However, it is informative to develop the analysis using the method in Table 5.4.1. We assume from the outset that our aim is to calculate $\Delta_r C_p^\infty$ and $\Delta_r H^\infty(\theta)$ at reference temperature θ.

We illustrate application of the Valentiner equation (Table 5.4.3) using data describing the dependence on temperature of the acid dissociation constant for ethanoic acid(aq) [11] (Table 5.4.4). The calculated of $\Delta_r C_p^\infty$ is close to that determined calorimetrically [14], $-(143 \pm 5)\,\mathrm{J\,K^{-1}\,mol^{-1}}$. At 298.2 K, the limiting enthalpy of dissociation is small, $-1.25\,\mathrm{J\,mol^{-1}}$. The dependence of $\Delta_r H^\infty(\mathrm{aq})$ on temperature is linear (Fig. 5.4.1), passing through zero near 295 K. Table 5.4.4 records the overall standard error and the estimates of a_1, a_2 and a_3 together with their standard errors. The latter are calculated using the diagonal elements of the variance–covariance matrix. The (symmetric) correlation matrix is shown in Table 5.4.5, the magnitude of each element being close to unity. The factors determining the signs and magnitudes of the off-diagonal elements are complicated. (The diagonal elements c_{jj} are unity and the magnitudes of all other elements c_{ij} are less than 1.0.) In general terms correlation coefficients reflect the structure of the data and the structure of the equations used to describe the dependence of, in this case, $\ln K(T)$ on temperature. There is little one can do about the data but it is worthwhile looking at the structure of the equation. The magnitude of the correlation coefficient c_{ij} measures the interdependence of coefficients a_i and a_j. If the magnitude of c_{ij} is close to unity this dependence is strong. In other words the error in the least square estimate of a_i is linked with the error in the least squares estimate of a_j. In the case considered here, this link is very strong. One can understand the reasons by examining the independent variables $(1/T)$ and $\ln(T/\mathrm{K})$. As shown in Table 5.4.6, these variables are close to being linearly interdependent over a small temperature range. Strong correlation is not a unique feature of the Valentiner equation. Many analytical procedures attempt to uncouple errors in the estimates. The root of the problem lies in the fact that the raw data involve just one independent variable, temperature. By measuring the dependence on temperature of one property, $\ln K(T)$ we hope to be in a position to calculate several thermodynamic variables.

A common theme in this and the next chapter concerns calculation of isobaric heat capacities of both reaction and activation. Many interesting procedures have been

Table 5.4.3. Valentiner Equation

At fixed pressure p.
Assumption: $\Delta_r C_p^\infty$ = independent of temperature
Then at temperature T and θ, $\Delta_r H^\infty(T) = \Delta_r H^\infty(\theta) + \Delta_r C_p^\infty(T - \theta)$
or $\Delta_r H^\infty(T)/T^2 = \Delta_r H^\infty(\theta)/T^2 + \Delta_r C_p^\infty(1/T)[1 - (\theta/T)]$
With $\theta/\mathrm{K} = 1$ and $T \simeq$ ambient temperature and above

$$1 - (\theta/T) \simeq 1.0$$

Then, $\Delta_r H^\infty(T)/T^2 = \Delta_r H^\infty(\theta)/T^2 + \Delta_r C_p^\infty(1/T)$

$$\int_\theta^T \mathrm{d}\ln K = \int_\theta^T \{[\Delta_r H^\infty(\theta)/RT^2] + [\Delta_r C_p^\infty/R](1/T)]\}\mathrm{d}T$$

Hence, $\ln K(T) =$

$$\ln K(\theta) + [[\Delta_r H^\infty(\theta)/R](-1/T) + [\Delta_r C_p^\infty/R]\ln T]_\theta^T$$

Therefore, $\ln K(T) = \{\ln K(\theta) + [\Delta_r H^\infty(\theta/\mathrm{K} = 1)/R\mathrm{K}]\} - [\Delta_r H^\infty(\theta/\mathrm{K} = 1)/R](1/T)$
$+ [\Delta_r C_p^\infty/R]\ln(T/\mathrm{K})$
This equation has the form $\ln K(T) = a_1 + (a_2 K/\mathrm{T}) + a_3 \ln(T/\mathrm{K})$
where $a_1 = \ln K(\theta) + [\Delta_r H^\infty(\theta/\mathrm{K} = 1)/R\mathrm{K}]$

$$a_2 = -\Delta_r H^\infty(\theta/\mathrm{K} = 1)/R\mathrm{K}], \text{ and } a_3 = \Delta_r C_p^\infty/R$$

limit$(T \to 0)\ln K(\mathrm{sln})$ is indeterminate
limit$(T \to \infty)\ln K(\mathrm{sln})$ is indeterminate
Also $\Delta_r H^\infty(T) = -a_2 R + a_3 R(T - \theta) = R[a_3(T - \theta) - (a_2\mathrm{K})]$
Since $-RT\ln K(T) = \Delta_r H^\infty(T) - T\Delta_r S(\mathrm{sln}; m_j = 1; \gamma_j = 1)$
then $-RT[a_1 + (a_2\mathrm{K}/T) + a_3\ln(T/\mathrm{K})] = R[a_3(T - \theta) - (a_2\mathrm{K})]$
$- T\Delta_r S(T; \mathrm{sln}; m_j = 1; \gamma_j = 1)$
or $RTa_1 + RTa_3\ln(T/\mathrm{K})] - Ra_3(T - \theta) = T\Delta_r S(T; \mathrm{sln}; m_j = 1; \gamma_j = 1)$
Hence $\Delta_r S(T; \mathrm{sln}; m_j = 1; \gamma_j = 1) = R[a_1 + a_3\ln(T/\mathrm{K})] - a_3\{1 - (\theta/T)\}$
But again, $1 - (\theta/T) \simeq 1.0$
$\Delta_r S(T; \mathrm{sln}; m_j = 1; \gamma_j = 1) = R[a_1 + a_3\{1 + \ln(T/\mathrm{K})\}]$
Then, $T\mathrm{d}\Delta_r S(\mathrm{sln}; m_j = 1; \gamma_j = 1)/\mathrm{d}T = TR[a_3/T] = Ra_3 = \Delta_r C_p^\infty/T$

Table 5.4.4. Ethanoic acid(aq)

Ambient pressure
Reference [13]; Temperature range: $273.15 \leqslant T/\mathrm{K} \leqslant 303.15$

$$K(\mathrm{aq}; 298.15; \mathrm{expt}) = 1.754 \times 10^{-5}$$

$$a_1 = 97.47 \pm 12.00;\ a_2 = -(4.812 \pm 0.518) \times 10^3;\ a_3 = -16.19 \pm 1.80$$

At 298.15 K, $\Delta_r H^\infty(\mathrm{aq})/\mathrm{J\,mol^{-1}} = -1.25$; $\Delta_r C_p^\infty(\mathrm{aq})/\mathrm{J\,K^{-1}\,mol^{-1}} = -135$

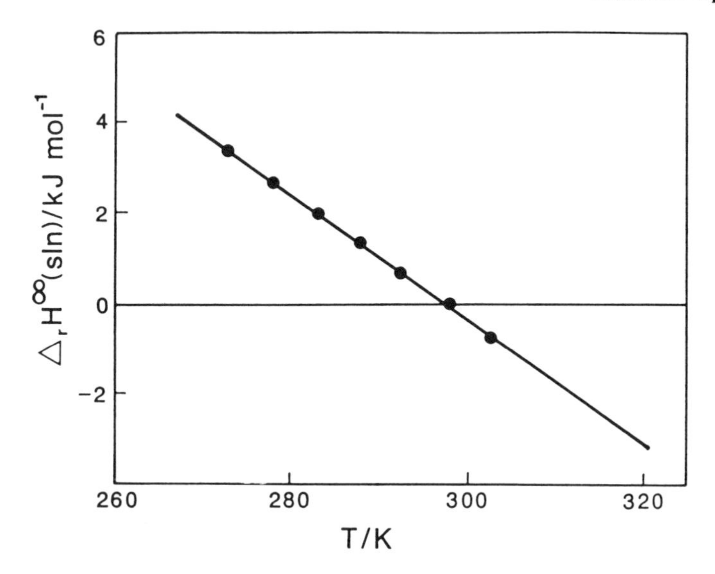

Fig. 5.4.1. Dependence on temperature of the limiting enthalpy of dissociation of ethanoic acid(aq) (see Table 5.4.4).

Table 5.4.5. Valentiner equation—ethanoic acid(aq)

Correlation matrix (see Table 5.4.4)

	1	2	3
1	1.0000000		
2	−0.9999183	1.0000000	
3	−0.9999975	0.9998869	1.0000000

Table 5.4.6. Temperature functions

$273.15 \leqslant T/\text{K} \leqslant 303.15$

(a) Equation: $(\text{K}/T) = a_1 + a_2 T/\text{K}$

where $a_1 = (6.952 \pm 0.046) \times 10^{-3}$, $a_2 = -(1.2069 \pm 0.016) \times 10^{-7}$,
standard error $= 4.297 \times 10^{-6}$

(B) Equation: $\ln(\text{K}/T) = a_1 + a_2 T/\text{K}$

where $a_1 = 4.6622 \pm 0.0067$, $a_2 = 3.473 \pm 0.0233$, standard error $= 6.180 \times 10^{-4}$

suggested which offer a rapid estimate of $\Delta_r C_p^{\infty}$ (and $\Delta^{\neq} C_p^{\infty}$) from the data. A simple graphical method for the estimation of a temperature-independent $\Delta_r C_p^{\infty}$ is useful. One approach [10] has merit in that it is based on an equation for dependence of the product, $T \ln K$ on temperature (Table 5.4.7). An application of this method is

Table 5.4.7. $T \ln K$ as a function of temperature, T

From Table 5.4.3, $\ln K(\text{aq}; T) = a_1 + a_2\text{K}/T + a_3 \ln(T/\text{K})$
Then, $T \ln K(\text{aq}; T) = a_1 T + a_2\text{K} + a_3 T \ln(T/\text{K})$
$\mathrm{d}[T \ln K]/\mathrm{d}T] = a_1 + a_3 + a_3 \ln(T/\text{K})$
Writing this equation as a difference function:

$$\Delta(T \ln K) = b + a_3 \ln(T/\text{K}) \quad \text{(a)}$$

where $\Delta_r C_p^\infty(\text{sln}) = a_3 R$

Thus an estimate of $\Delta_r C_p^\infty$ is obtained from the dependence of $\Delta(T \ln K)/\Delta T$ on $\ln(T/\text{K})$

$$\text{Thus knowing } a_3, [\ln K - a_3 \ln(T/\text{K})] = a_1 + a_2\text{K}/T \quad \text{(b)}$$

A linear least squares analysis yields parameters a_1 and a_2.

Table 5.4.8. Methanoic acid(aq)

Ambient pressure.
Reference [15]: Temperature range $273.15 \leqslant T/\text{K} \leqslant 333.15$
Equation (a) of Table 5.4.7:

$$a_1 = 115.4 \pm 46.3,\ a_2 = -22.23 \pm 8.11,\ \text{standard error} = 1.62$$

$$\Delta_r C_p^\infty(\text{aq})/\text{J K}^{-1}\text{mol}^{-1} = -185.$$

Equation (b) of Table 5.4.7:

$$a_1 = 137.964 \pm 0.07,\ a_2 = -(6.6314 \pm 0.0021) \times 10^3,$$

$$\Delta_r H^\infty(\text{aq}; 298.2\ \text{K})/\text{J mol}^{-1} = 28.94$$

---

reported in Table 5.4.8 with respect to the dependence on temperature of the acid dissociation constant for methanoic acid(aq) [15].

The Valentiner equation also has been used in conjunction with the solubilities in, mainly, aqueous solutions. Here an equilibrium involving X(g) and X(aq) for chemical substance X is characterized by a set of thermodynamic parameters including $\Delta_{\text{sin}} C_p^\infty$ for the process X(g) → X(aq). The Valentiner equation has been used to describe the dependences on temperature of solubilities in water for methyl halides [16], mercury [17], and naphthalene [18], and of the critical micellar concentration (cmc) for surfactants in aqueous solutions [19].

Chemical kinetics

For the majority of kinetic data, the analysis described in Table 5.3.3 is adequate. However in many cases such is not the case [20–24] as is often revealed in a preliminary inspection of the data using the integrated form of the Arrhenius equation (Table 5.4.9). A mean activation energy is calculated from rate constants at pairs of tempera-

Table 5.4.9. Inspection of kinetic data

At fixed pressure.
Arrhenius equation: $\ln k^{\#} = a_1 - \Delta E/RT$
Then $\ln[k^{\#}(T_2)/k^{\#}(T_1)] = -(\Delta E/R)(T_2^{-1} - T_1^{-1}) = -(\Delta E/R)(T_1 - T_2)/(T_1 T_2)$
or $\Delta E = \{RT_1 T_2/(T_2 - T_1)\} \ln[k^{\#}(T_2)/k^{\#}(T_1)]$

Table 5.4.10. Chemical kinetics—analysis using reference temperature

Assumption: $\Delta^{\neq} C_{\mathrm{p}}^{\infty}(\mathrm{sln})$ = constant at fixed pressure;

$$\Delta^{\neq} H^{\infty}(T) = \Delta^{\neq} H^{\infty}(\theta) + \Delta^{\neq} C_{\mathrm{p}}^{\infty}(T - \theta)$$

Therefore,

$$\ln[k^{\#}(T)\mathrm{K}/T] = \ln[k^{\#}(\theta)\mathrm{K}/\theta] + [\Delta^{\neq} H^{\infty}(\theta)/R][(1/\theta) - (1/T)] + [\Delta^{\neq} C_{\mathrm{p}}^{\infty}/R][\ln(T/\theta) + (\theta/T) - 1]$$

which has the general form,

$$\ln[k^{\#}(T)\mathrm{K}/T] = a_1 + a_2[(\mathrm{K}/\theta) - (\mathrm{K}/T)] + a_3[\ln(T/\theta) + (\theta/T) - 1]$$

where $a_1 = \ln[k^{\#}(\theta)\mathrm{K}/\theta]$, $a_2 = [\Delta^{\neq} H^{\infty}(\theta)/R\mathrm{K}]$, $a_3 = [\Delta^{\neq} C_{\mathrm{p}}^{\infty}/R]$

tures T_1 and T_2 across the experimental temperature range. Clear indication that the limiting isobaric heat capacity of activation, $\Delta^{\neq} C_{\mathrm{p}}^{\infty}(\mathrm{s\,ln}; T; p)$ is not negligibly small is shown by a marked curvature in a plot of $E(T_{\mathrm{m}})$ against temperature T_{m} where T_{m} is the arithmetic mean of T_1 and T_2.

The starting point is the treatment set out in Table 5.4.10 in which we assume that the limiting enthalpy of activation $\Delta^{\neq} H^{\infty}(T)$ is a linear function of $T - \theta$ where θ is a reference temperature. The result is the 'Valentiner equation with reference'. In effect the equation expresses the dependence of $\ln[k^{\#}(T)\mathrm{K}/\mathrm{T}]$ on temperature using three parameters in which a_3 measures the limiting isobaric heat capacity of activation $\Delta^{\neq} C_{\mathrm{p}}^{\infty}$. Because we have based the analysis on transition state theory, it is convenient to express the dependence on temperature of the variable $\ln[k^{\#}(T)\mathrm{K}/T]$ rather than $\ln[k^{\#}(T)]$. An example of the analysis (Table 5.4.10) is shown in Table 5.4.11. The dependence on temperature of the percentage difference between $\ln[k^{\#}(\mathrm{obs})\mathrm{K}/T]$ and $\ln[k^{\#}(\mathrm{calc})\mathrm{K}/T]$ shows no pattern (Fig. 5.4.2) indicating that the equation fits the data satisfactorily. In other words there is no hint of a meaningful dependence on temperature of $\Delta^{\neq} C_{\mathrm{p}}^{\infty}(\mathrm{aq})$. By choosing a reference temperature near the middle of the experimental range we have produced a correlation matrix (Table 5.4.12) in which the off-diagonal elements differ considerably from unity.

For aqueous solutions the approximation which sets $1 - (\theta/T) \simeq 1.0$ is valid if $\theta/k = 1.0$ (Table 5.4.13). Then $\ln[k^{\#}(T)\mathrm{K}/T]$ is expressed as a function of three variables (Table 5.4.14), one form of the Valentiner equation as applied to kinetic

Table 5.4.11. 2-Chloro-2-methylpropane(aq)

Reference [26]: Temperature range: $274.158 \leqslant T/\text{K} \leqslant 293.163$
Reference temperature: $\theta/\text{K} = 283.173$

$$a_1 = -5.636612 \pm 0.00064, \quad a_2 = (1.22588 \pm 0.00059) \times 10^4,$$

$$a_3 = -41.07 \pm 2.3$$

$\Delta^{\#}H^{\infty}(\text{aq})/\text{kJ mol}^{-1} = 101.92$, $\Delta^{\#}C_p^{\infty}(\theta)/\text{J K}^{-1}\text{mol}^{-1} = -341.5$

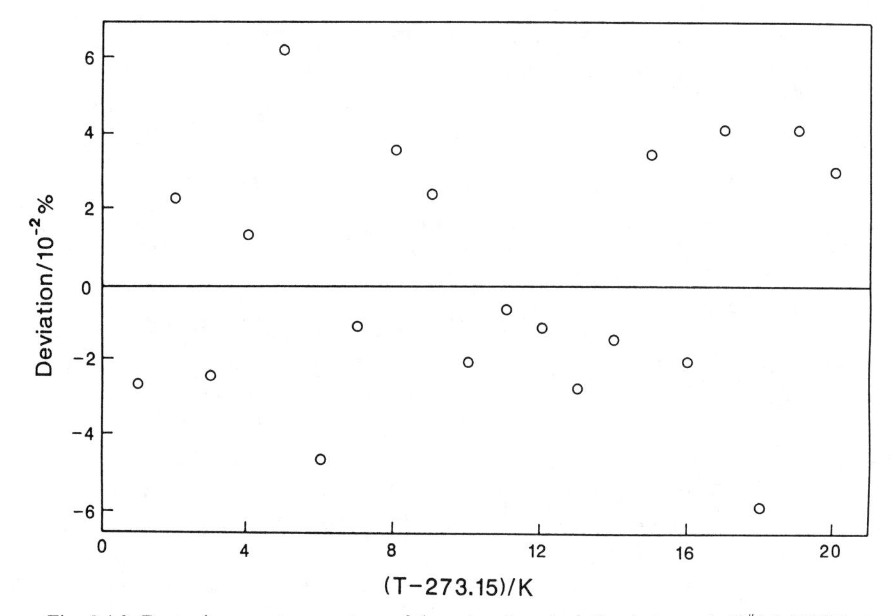

Fig. 5.4.2. Dependence on temperature of the percentage deviation between $\ln[k^{\#}(\text{obs})\text{K}/T]$ and $\ln[k^{\#}(\text{calc})\text{K}/T]$ for the hydrolysis of 2-chloro-2-methylpropane(aq) using the data in [26] and the equation in Table 5.4.10.

Table 5.4.12. Correlation matrix

Data set: see Table 5.4.11. Equation: see Table 5.4.10.

	1	2	3
1	1.0000000		
2	0.044	1.0000000	
3	−0.745	−0.116	1.0000000

Table 5.4.13. Valentiner Equation—Kinetics

At fixed pressure, p.
Assumption $\Delta^{\neq} C_{p}^{\infty}(\text{sln})$ = independent of temperature
Then $\Delta^{\neq} H^{\infty}(T) = \Delta^{\neq} H^{\infty}(\theta) + \Delta^{\neq} C_{p}^{\infty}(T - \theta)$
or $\Delta^{\neq} H^{\infty}(T)/T^2 = \Delta^{\neq} H^{\infty}(\theta)/T^2 + \Delta^{\neq} C_{p}^{\infty}(1/T)[1 - (\theta/T)]$

Put $\theta/\text{K} = 1$. Then at ambient temperature, $1 - (\theta/T) \simeq 1.0$
Hence, $\ln[k^{\neq}(T)\text{K}/T] - \ln k^{\neq}(\theta)K/\theta]$

$$[\Delta_r H(\theta)/R][-(1/T) + (1/\theta)] + [\Delta^{\neq} C_{p}^{\infty}/R]\ln(T/\theta)$$

or with $\theta/\text{K} = 1$,

$$\ln[k^{\neq}(T)\text{K}/T] = \{\ln[k^{\neq}(\theta/\text{K} = 1)] + \Delta_r H(\theta/\text{K} = 1)/R\text{K}\} - [\Delta_r H(\theta/\text{K} = 1)/R](1/T) + [\Delta^{\neq} C_{p}^{\infty}/R]\ln(T/\text{K})$$

which has the form
$\ln[k^{\neq}(T)\text{K}/T] = a_1 + a_2(\text{K}/T) + a_3\ln(T/\text{K})$
where $a_1 = \ln[k^{\neq}(\theta/\text{K} = 1)] + \Delta_r H(\theta/\text{K} = 1)/R\text{K}$,
$a_2 = -\Delta^{\neq} H^{\infty}(\theta/\text{K} = 1)/R\text{K}, \quad a_3 = \Delta^{\neq} C_{p}^{\infty}/R$
Hence $\Delta^{\neq} H^{\infty}(T) = \Delta^{\neq} H^{\infty}(\theta/\text{K} = 1) + \Delta^{\neq} C_{p}^{\infty}[T - \theta]$

$$\Delta^{\neq} H^{\infty}(T) = -a_2 R\text{K} + a_3 R[T - \theta] = R[a_3(T - \theta) + a_2\text{K}]$$

Table 5.4.14. Valentiner equation—chemical kinetics (at pressure p)

$\ln k^{\#} = a_1 + a_2\text{K}/T + a_3\ln(T/\text{K})$.

$$\text{d}\ln k^{\#}/\text{d}T = -a_2\text{K}/T^2 + a_3/T$$

(see Table 5.1.1) $\Delta^{\neq} H^{\infty}(T) = RT^2[\text{d}\ln k^{\#}/\text{d}T - (1/T)]$
$= RT^2[-a_2\text{K}/T^2 + a_3/T - 1/T]$
$= R[-a_2\text{K} + a_3 T - T]$

Hence*, $\Delta^{\neq} H^{\infty}(T/\text{K} = 1) = -a_2 R\text{K}$

$$\Delta^{\neq} H^{\infty}(T/\text{K}) = R[-a_2\text{K} + T(a_3 - 1)]$$

and $\Delta^{\neq} C_{p}^{\infty}/R = a_3 - 1$

*Units: $[\text{J mol}^{-1}]/[\text{J mol}^{-1}\text{ K}^{-1}] = [1][\text{K}] + [\text{K}][1]$

data. By way of an example we apply the equation to kinetic data [25] describing the solvolysis of benzyl chloride(aq) (Table 5.4.15). The scatter in the dependence on temperature of the percentage difference between $\ln[k^{\#}(\text{obs})\text{K}/T]$ and $\ln[k^{\#}(\text{calc})\text{K}/T]$ is satisfactory (Fig. 5.4.3). The Valentiner equation has been used by many authors to analyse kinetic data [27–36], usually for solvolytic reactions in aqueous solutions

Table 5.4.15. Benzyl chloride(aq)

Reference [21]; Temperature range: $288.118 \leqslant T/\mathrm{K} \leqslant 338.163$
$a_1 = 156.17 \pm 6.4$ $a_2 = -(168.79 \pm 2.97) \times 10^2$ $a_3 = -19.44 \pm 0.95$

$\Delta^{\neq} H^{\infty}(\mathrm{aq}; 323.138\ \mathrm{K})/\mathrm{kJ\ mol^{-1}} = 88.26$

$\Delta^{\neq} C_p^{\infty}/\mathrm{J\ K^{-1}\ mol^{-1}} = -161.7$

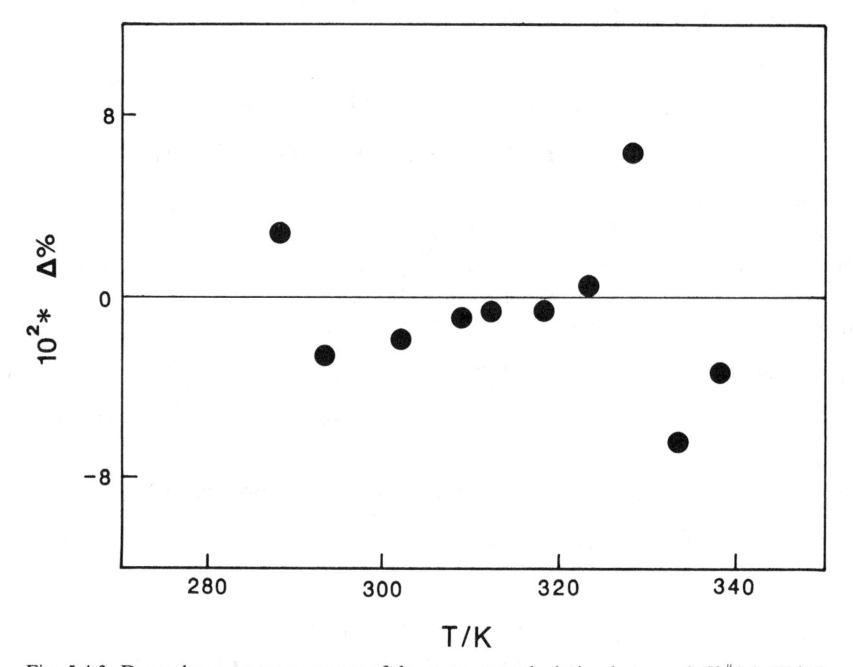

Fig. 5.4.3. Dependence on temperature of the percentage deviation between $\ln[k^{\#}(\mathrm{obs})\mathrm{K}/T]$ and $\ln[k^{\#}(\mathrm{calc})\mathrm{K}/T]$ for the hydrolysis of benzyl chloride(aq) using the data in [27] and the equation in Table 5.4.15.

following the lead given by Moelwyn-Hughes [34]. For the most part these treatments are based on the equation given in Table 5.4.13. In such instances strong correlation between errors on the estimates of derived parameters is observed [36]. The magnitudes of the off-diagonal correlation coefficients are diminished by using a reference temperature (see Table 5.4.12) near the middle of the experimental range.

The Valentiner equation has been extended by adding further terms, e.g. a term linear in temperature [38–40].

References to section 5.4

[1] S. Valentiner. *Z. Physik.*, 1927, **42**, 253.
[2] D. H. Everett; W. F. K. Wynne-Jones. *Trans. Faraday Soc.*, 1939, **37**, 1380.

[3] D. H. Everett; W. F. K. Wynne-Jones. *Proc. Roy. Soc. A*, 1938, **169**, 190; 1941, **177**, 499.
[4] D. H. Everett; D. A. Landsman; B. R. W. Pinsent. *Proc. Roy Soc. A*, 1952, **215**, 403.
[5] M. C. Cox; D. H. Everett; D. Landsman; R. J. Munn. *J. Chem. Soc. B*, 1968, 1373.
[6] B. A. Timimi; D. H. Everett. *J. Chem. Soc. B*, 1968, 1380.
[7] E. A. Moelwyn-Hughes. *Proc. Roy. Soc. A*, 1938, **164**, 295.
[8] D. N. Glew; E. A. Moelwyn-Hughes. *Proc. Roy. Soc. A*, 1952, **211**, 254.
[9] W. van der Linde; D. Northcott; W. Redmond; R. E. Robertson. *Canad. J. Chem.*, 1969, **47**, 279.
[10] R. W. Ramette, *J. Chem. Educ.*, 1977, **54**, 280.
[11] S. Bergstrom; G. Olofsson. *J. Soln. Chem.*, 1975, **4**, 535.
[12] D. A. Palmer; R. W. Ramette; R. E. Mesmer. *J. Soln. Chem.*, 1984, **13**, 673.
[13] H. S. Harned; R. W. Ehlers. *J. Amer. Chem. Soc.*, 1932, **54**, 1350.
[14] G. Olofsson. *J. Chem. Thermodynamics*, 1984, **16**, 39.
[15] H. S. Harned; N. D. Embree. *J. Amer. Chem. Soc.*, 1934, **56**, 1042.
[16] D. N. Glew; E. A. Moelwyn-Hughes, *Disc. Faraday Soc.*, 1953, **15**, 150.
[17] D. N. Glew; D. A. Hames. *Canad. J. Chem.*, 1971, **49**, 3114; 1972, **50**, 3124.
[18] P. Perez-Tejeda; C. Yanes; A. Maestre. *J. Chem. Eng. Data*, 1990, **35**, 244.
[19] N. Nishikido. *Langmuir*, 1990, **6**, 1225.
[20] R. E. Robertson. *Suomen Kemistilehti A*, 1960, **33**, 63.
[21] R. E. Robertson. *Progr. Phys. Org. Chem.*, 1967, **4**, 213.
[22] M. J. Blandamer; R. E. Robertson; J. M. W. Scott. *Prog. Phys. Org. Chem.*, 1985, **15**, 149.
[23] G. Kohnstam. *Adv. Phys. Org. Chem.*, 1967, **5**, 121.
[24] M. J. Blandamer, H. S. Golinkin; R. E. Robertson. *J. Amer. Chem. Soc.*, 1969, **91**, 2678.
[25] S. Wold. *Acta. Chem. Scand.*, 1970, **24**, 2321.
[26] E. A. Moelwyn-Hughes; R. E. Robertson; S. E. Sugamori. *J. Chem. Soc.*, 1965, 1965.
[27] R. E. Robertson; J. M. W. Scott. *J. Chem. Soc.*, 1961, 1596.
[28] A. Queen. *Canad. J. Chem.*, 1967, **45**, 1619.
[29] N. J. Cleve; E. K. Euranto. *Suomen Kemistilehti, B*, 1964, **37**, 126.
[30] N. J. Cleve. *Suomen Kemistilehti, B*, 1972, **45**, 235; 1972, **46**, 385.
[31] P. A. Adams; J. G. Sheppard. *J. Chem. Soc. Faraday Trans. 1*, 1980, **76**, 2114.
[32] J. Biordi; E. A. Moelwyn-Hughes. *J. Chem. Soc.*, 1962, 4291.
[33] I. Fells; E. A. Moelwyn-Hughes. *J. Chem. Soc.*, 1959, 398.
[34] R. L. Heppolette; R. E. Robertson. *Proc. Roy. Soc.*, 1959, **252A**, 273.
[35] M. Tonnet; A. N. Hambly. *Aust. J. Chem.*, 1970, **23**, 2427.
[36] R. Huq. *J. Chem. Soc. Faraday Trans. 1*, 1973, **69**, 98.
[37] E. A. Moelwyn-Hughes. *Proc. Roy. Soc.*, 1938, **A164**, 295.
[38] M. J. Blandamer; R. E. Robertson; J. M. W. Scott. *Canad. J. Chem.*, 1980, **58**, 772.
[39] B. Rossall; R. E. Robertson. *Canad. J. Chem.*, 1975, **53**, 869.
[40] A. Marchetti; E. Picchioni; L. Tassi; and G. Tosi. *Analyt. Chem.*, 1989, **61**, 1971.

5.5 ROBINSON EQUATION

Chemical equilibria

In the previous section we started out with the assumption that, for a given reaction, $\Delta_r C_p^\infty$ is independent of temperature. Although this assumption is unsatisfactory, we often have no information on which to base an alternative treatment. The only recourse is to assume a particular dependence. One procedure assumes that $\Delta_r C_p^\infty$ is given by the product of the partial derivative $(\partial \Delta_r C_p^\infty / \partial T)$ and temperature and that $(\partial \Delta_r C_p^\infty / \partial T)$ is independent of temperature. The consequences of these assumptions are explored in Table 5.5.1 in which two integrations yield an equation for the dependence on temperature of $\ln K(T)$ with the proviso that $\Delta_r C_p^\infty$ is not zero. The final equation uses two temperature functions, $[(1/\theta) - (1/T)]$ and $[T + (\theta^2/T) - 2\theta]$ to yield in a least squares analysis estimates of three quantities: $\ln K(\theta)$, $\Delta_r H^\infty(\theta)$ and $(\partial \Delta_r C_p^\infty / \partial T)$. The assumption in the Robinson Equation [1–3] is that $(\partial \Delta_r C_p^\infty / \partial T)$ has statistical significance. The isobaric heat capacity of reaction does not itself emerge from the least squares analysis. Rather it is equal to the product, $T(\partial \Delta_r C_p^\infty / \partial T)$.

Table 5.5.1. Robinson equation

At fixed pressure.
Assumption: $\Delta_r C_p^\infty = T\{\partial \Delta_r C_p^\infty / \partial T\}$,
where $\partial \Delta_r C_p^\infty / \partial T$ = independent of temperature
With reference temperature, θ using equation (ii) of Table 5.2.2

$$\Delta_r H^\infty(T) = \Delta_r H^\infty(\theta) + (\partial \Delta_r C_p^\infty / \partial T) \int_\theta^T T \mathrm{d}T$$

Hence, $\Delta_r H^\infty(T) = \Delta_r H^\infty(\theta) + (1/2)(\partial \Delta_r C_p^\infty / \partial T)(T^2 - \theta^2)$
From the van't Hoff equation (equation (i) of Table 5.2.2),

$$\ln K(T) = \ln K(\theta) + \int_\theta^T [[\Delta_r H^\infty(\theta)/RT^2] + (1/2R) \times (\partial \Delta_r C_p^\infty / \partial T)[1 - (\theta^2/T^2)]]\mathrm{d}T$$

or

$$\ln K(T) = \ln K(\theta) + [\Delta_r H^\infty(\theta)/R][-1/T]_\theta^T + (1/2R)(\partial \Delta_r C_p^\infty / \partial T)[T + (\theta^2/T)]_\theta^T$$

Then,

$$\ln K(T) = \ln K(\theta) + \{[\Delta_r H^\infty(\theta)/R][(1/\theta) - (1/T)]\} + \{(1/2R)(\partial \Delta_r C_p^\infty / \partial T)[T + (\theta^2/T) - 2\theta]\}$$

Table 5.5.2. Robinson equation—conventional approach

From Table 5.5.1,

$$\ln K(T) = \{\ln \mathrm{K}(\theta) + [\Delta_r H^{\infty}(\theta)/R\theta] - (\theta/R)[\partial \Delta_r C_p^{\infty}/\partial T]\}$$
$$+ \{-[\Delta_r H^{\infty}(\theta)/R\mathrm{K}] + (\theta^2/2R\mathrm{K})[\partial \Delta_r C_p^{\infty}/\partial T]\}\{\mathrm{K}/T\}$$
$$+ \{(\mathrm{K}/2R)[\partial \Delta_r C_p^{\infty}/\partial T]\}(T/\mathrm{K})$$

With $\theta/\mathrm{K} = 1$, $\ln K(T) = a_1 + a_2 \mathrm{K}/T + a_3 T/\mathrm{K}$
where

$$a_1 = \ln K(\theta/\mathrm{K} = 1) + [\Delta_r H^{\infty}(\theta/\mathrm{K} = 1)/R\mathrm{K}] - (\mathrm{K}/R)[\partial \Delta_r C_p^{\infty}/\partial T]$$
$$a_2 = -[\Delta_r H^{\infty}(\theta/\mathrm{K} = 1)/R\mathrm{K}] + (\mathrm{K}/2R)[\partial \Delta_r C_p^{\infty}/\partial T]$$
$$a_3 = (\mathrm{K}/2R)[\partial \Delta_r C_p^{\infty}/\partial T]$$

Table 5.5.3. Robinson equation—chemical equilibria

Conventional equation:

$$\ln K = a_1 + a_2(T/\mathrm{K})^{-1} + a_3(T/\mathrm{K})$$
$$\mathrm{d} \ln K/\mathrm{d}T = -a_2(\mathrm{K}/T^2) + a_3/K.$$
$$\mathrm{d}^2 \ln K/\mathrm{d}T^2 = 2a_2 \mathrm{K}/T^3$$

At extremum in $\ln K$, $T/\mathrm{K} = (a_2/a_3)^{1/2}$
At temperature T; $\Delta_r H^{\infty}(T) = -Ra_2 \mathrm{K} + Ra_3(T^2/\mathrm{K})$

$$\Delta_r C_p^{\infty}(T) = 2a_3 RT/\mathrm{K}$$

and $\mathrm{d}\Delta_r C_p^{\infty}/\mathrm{d}T = 2a_3 R/\mathrm{K}$

$$\Delta_r S(\mathrm{sln}; m_j = 1; \gamma_j = 1; T) = R(a_1 + 2a_3 T/\mathrm{K})$$

and $\mathrm{d}\Delta_r S(\mathrm{sln}; m_j = 1; \gamma_j = 1; T)/\mathrm{d}T = 2a_3 R/\mathrm{K} = \Delta_r C_p^{\infty}/T$
limit$(T \to 0) \ln K$ is indeterminate; limit$(T \to \infty) \ln K$ is indeterminate
At $T/\mathrm{K} = 1$, $\ln K = a_1 + a_2 + a_3$

The conventional form of the Robinson equation is based on a reference temperature $\theta/K = 1$ (Table 5.5.2). The derived equation in Table 5.5.2 can be written in the simple form shown in Table 5.5.3. The Robinson Equation can account for extrema in the dependence of $\ln K(T)$ on T. The Robinson equation retains the classic (K/T)-term but a $\ln(T/\mathrm{K})$ term is missing, being replaced by a term which is linear in temperature. Interestingly the dependence of $\ln K(T)$ on temperature shows indeterminate limits at both zero and infinite temperatures. A plot of $\ln K(T)$ does however pass though the point $(a_1 + a_2 + a_3)$ at $T/\mathrm{K} = 1.0$. An example of the

Table 5.5.4. 2,2,2-Trideuteroethanoic acid(aq)

Reference [3]. Temperature range: $273.15 \leqslant T/\mathrm{K} \leqslant 323.15$
$a_1 = 5.907 \pm 0.53$; $a_2 = -(2.5019 \pm 0.0078) \times 10^3$; $a_3 = -(2.85 \pm 0.089) \times 10^{-2}$

st. error $= 2.193 \times 10^{-3}$

$K(\text{expt; aq; 298.15 K}) = 1.690 \times 10^{-5}$; $\Delta_r H^\infty(\text{aq; 298.15 K})/\mathrm{J\,mol^{-1}} = -274.9$

$\Delta_r C_p^\infty(\text{aq; 298.15 K})/\mathrm{J\,K^{-1}\,mol^{-1}} = -141.4$

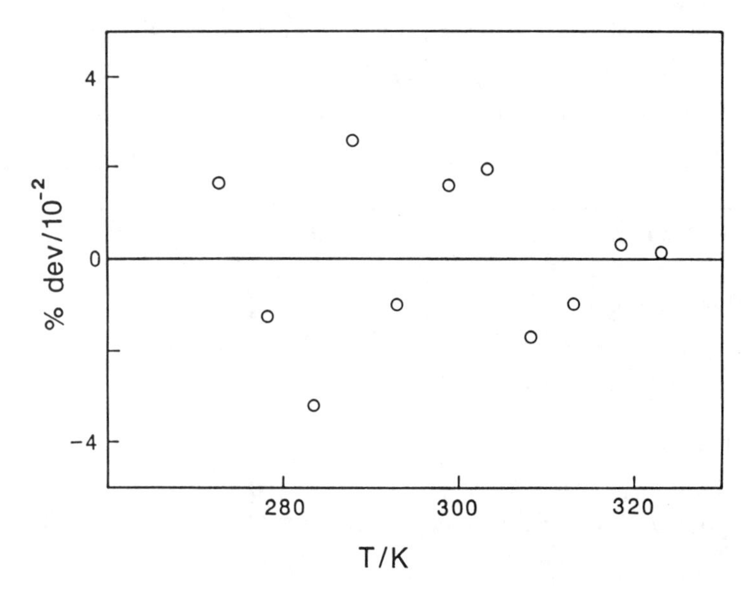

Fig. 5.5.1. Dependence on temperature of the percentage deviation between ln K(obs) and ln K(calc) for the dissociation of 2,2,2-trideuteroethanoic acid(aq); (see Table 5.5.4).

application of this equation is summarized in Table 5.5.4 with reference to the dependence on temperature of the acid dissociation constant [3] for 2,2,2-trideutero-ethanoic acid(aq). The dependence on temperature of the percentage difference between ln K(calc) and ln K(obs) shows a satisfactory scatter (Figure 5.5.1). The dependence on temperature of the limiting enthalpy of reaction, $\Delta_r H^\infty$(aq) is summarized in Fig. 5.5.2. The Robinson equation has been extensively used [4–14] to express the dependence on temperature of acid dissociation constants in aqueous solutions. A Robinson equation is also used to express the dependence on temperature of the basic equilibrium constant for ammonia in aqueous solutions [15]. Closely related to the Robinson equation is the procedure in which the Gibbs energy of reaction is expressed as a cubic in temperature T (Tables 5.5.5 and 5.5.6).

The Robinson equation (1) uses the first three terms (Table 5.5.3) of a polynomial in T/K. In fact many of the equations used to describe the dependence on temperature

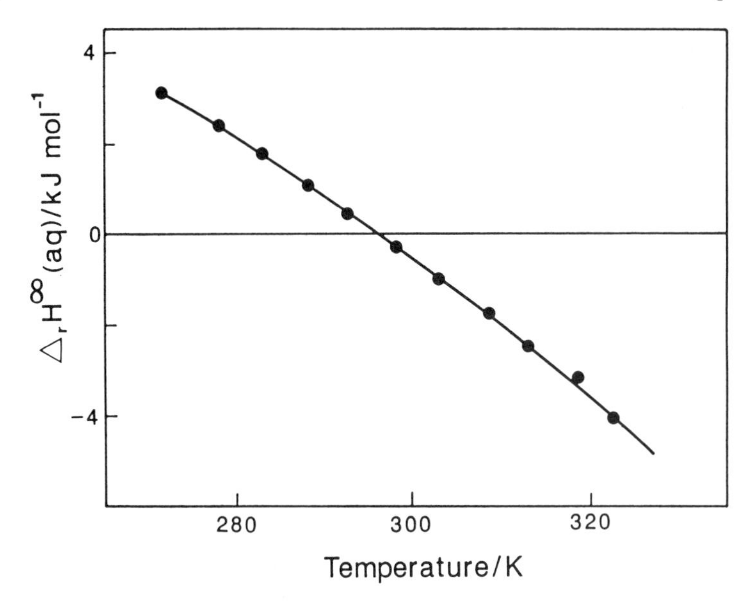

Fig. 5.5.2. Dependence on temperature of the limiting enthalpy of dissociation of 2,2,2-trideuteroethanoic acid(aq); (see Table 5.5.4).

Table 5.5.5. Dependence of Gibbs energy of reaction on temperature

Assuming $p \simeq p^0$

Equilibrium: $\Delta_r G^0(T)/R = a_1\mathrm{K} + a_2 T + a_3 T^2/\mathrm{K} + a_4(T^3/K^2)$

$$(1/R)[\Delta_r G^0(T)/T] = a_1(\mathrm{K}/T) + a_2 + a_3 T/\mathrm{K} + a_4 T^2/\mathrm{K}^2$$

$$(1/R)\mathrm{d}(\Delta_r G^0/T)/\mathrm{d}T = -a_1\mathrm{K}/T^2 + a_3/\mathrm{K} + 2a_4 T/\mathrm{K}^2$$
$$= -(1/R)\Delta_r H^\infty/T^2$$

Hence, $\Delta_r H^\infty(T)/R = a_1 K - a_3 T^2/\mathrm{K} - 2a_4 T^3/\mathrm{K}^2$

$$\Delta_r C_p^\infty(T) = R[-2a_3 T/\mathrm{K} - 6a_4 T^2/K^2]$$
$$\mathrm{d}\Delta_r C_p^\infty/\mathrm{d}T = R[-2a_3/\mathrm{K} - 12a_4 T/\mathrm{K}^2]$$

Table 5.5.6. Dependence of $\Delta_r G^0(T)$—quadratic equation

Equilibria: $\Delta_r G^0(T) = -RT\ln K(T)$

Equation: $\Delta_r G^0(T)/R = a_1\mathrm{K} + a_2 T + a_3 T^2/\mathrm{K}$

or $(1/R)[\Delta_r G^0(T)/T] = a_1 K/T + a_2 + a_3(T/\mathrm{K})$

$$(1/R)\mathrm{d}\{\Delta_r G^0(T)/T\}/\mathrm{d}T = -a_1\mathrm{K}/T^2 + a_3/\mathrm{K} = -(1/R)\Delta_r H^\infty/T^2$$

$$\Delta_r H^\infty(T) = -RT^2[-a_1\mathrm{K}/T^2 + a_3/\mathrm{K}]$$

Hence, $\Delta_r H^\infty(T) = R[a_1\mathrm{K} - a_3 T^2/\mathrm{K}]$, and $\Delta_r C_p^\infty(T) = -2a_3 RT/\mathrm{K}$

Table 5.5.7. Polynomial in T^{-1}

Equililbrium: $\ln K(T) = a_1 + \Sigma(j = 3; j = \mathrm{i})\{a_{j-1}/(j-2)\}[\mathrm{K}/T]^{(j-2)}$
Example $i = 5$

$$\ln K(T) = a_1 + a_2\mathrm{K}/T + (a_3/2)(\mathrm{K}/T)^2 + (a_4/3)(\mathrm{K}/T)^3$$

$$\mathrm{d}\ln K(T)/\mathrm{d}T = -a_2\mathrm{K}/T^2 - a_3\mathrm{K}^2/T^3 - a_4\mathrm{K}^3/T^4 = \Delta_r H^\infty(T)/RT^2$$

$$\Delta_r H^\infty(T) = -R[a_2\mathrm{K} + a_3\mathrm{K}^2/T + a_4\mathrm{K}^3/T^2]$$

of equilibrium constants can be linked through general polynomial equations. One procedure expresses $\ln K(T)$ in terms of a polynomial in (K/T) (Table 5.5.7).

References to section 5.5

[1] H. S. Harned; R. A. Robinson. *Trans. Faraday Soc.*, 1940, **36**, 973.
[2] R. Gary; R. G. Bates; R. A. Robinson. *J. Phys. Chem.*, 1965, **69**, 2750.
[3] M. Paabo; R. G. Bates; R. A. Robinson. *J. Phys. Chem.*, 1966, **70**, 540.
[4] M. Paabo; R. G. Bates; R. A. Robinson. *J. Phys. Chem.*, 1966, **70**, 2073.
[5] A. K. Covington; R. A. Robinson; R. G. Bates. *J. Phys. Chem.*, 1966, **70**, 3820.
[6] R. G. Bates; H. B. Hetzer. *J. Phys. Chem.*, 1961, **65**, 667.
[7] S. P. Datta; A. K. Grzybowski; B. A. Weston. *J. Chem. Sox.*, 1963, 792.
[8] R. N. Roy; R. A. Robinson; R. G. Bates. *J. Amer. Chem. Soc.*, 1973, **95**, 8231.
[9] S. Goldman; P. Sagner; R. G. Bates. *J. Phys. Chem.*, 1971, **75**, 826.
[10] R. A. Robinson; A. Peiperl. *J. Phys. Chem.*, 1963, **67**, 2860.
[11] J.-C. Halle; R. G. Bates. *J. Soln. Chem.*, 1975, **4**, 1033.
[12] B. D. Struck, M. Perec, D. Triefenbach. *J. Electroanal. Chem.*, 1986, **214**, 473.
[13] C. A. Vega, S. Delgado. *J. Chem. Eng. Data*, 1987, **32**, 218.
[14] S. L. Clegg, P. Brimblecombe. *J. Phys. Chem.*, 1989, **93**, 7237.

5.6 IVES AND PRYOR EQUATION

As a consequence of the underlying assumption, a feature of the Ives and Pryor equation [1] is the absence of a term in T^{-1} in the equation describing the dependence of $\ln K(T)$ on temperature (Table 5.6.1). Indeed we have to follow a rather tortuous path in order to obtain this equation starting from an assumption concerning isobaric heat capacities of reaction. We assume that the second differential $(\partial^2 \Delta_r C_p^\infty/\partial T^2)$ is a constant. Integration over the range θ to T yields an equation for $\Delta_r H^\infty(T)$ in terms of $\Delta_r H^\infty(\theta)$. Further we assume that $\Delta_r H^\infty(\theta)$ and $\Delta_r C_p^\infty(\theta)$ are zero at the reference temperature, $\theta/\mathrm{K} = 1.0$. Ives and Pryor used this equation (Table 5.6.2) to analyse the dependence on temperature of the acid dissociation constants for haloethanoic acids in aqueous solution (Table 5.6.3).

In general terms, the Ives and Pryor equation corresponds to the first three terms in a polynomial:

$$\ln K(T) = \Sigma(j = 1; j = 3)a_j(T/K)^{(j-1)} \tag{5.6.1}$$

Table 5.6.1. Ives and Pryor equation

At fixed pressure.
Assumption: $d^2\Delta_r C_p^\infty/dT^2$ = constant
Integration between temperature θ and temperature T.

$$\Delta_r C_p^\infty(T) = \Delta_r C_p^\infty(\theta) + (\partial\Delta_r C_p^\infty/\partial T)_\theta(T-\theta) + (\partial^2\Delta_r C_p^\infty/\partial T^2)_\theta(T-\theta)^2/2$$

A further integration:

$$\Delta_r H^\infty(T) = \Delta_r H^\infty(\theta) + \Delta_r C_p^\infty(\theta)(T-\theta) + [(\partial\Delta_r C_p^\infty/\partial T)_\theta(T-\theta)^2/2] + [(\partial^2\Delta_r C_p^\infty/\partial T^2)_\theta(T-\theta)^3/6]$$

We set the reference temperature, $\theta/\mathrm{K} = 1$
Assumptions: $\Delta_r H^\infty(\theta/\mathrm{K}=1) = 0$; and $\Delta_r C_p^\infty(\theta/\mathrm{K}=1) = 0$
At ambient temperatures $T \gg \theta$
Hence, $\Delta_r H^\infty(T) = (\partial\Delta_r C_p^\infty/\partial T)_\theta T^2/2 + (\partial^2\Delta_r C_p^\infty/\partial T^2)_\theta T^3/6$
From the van't Hoff equation,

$$\int_{\theta=1}^{T} d\ln K = \int_{\theta}^{T} [(1/2R)(\partial\Delta_r C_p^\infty/\partial T)_\theta + (T/6R)(\partial^2\Delta_r C_p^\infty/\partial T^2)_\theta]\, dT$$

or $\ln K(T) = \ln K(\theta/\mathrm{K}=1) + (\mathrm{K}/2R)(\partial\Delta_r C_p^\infty/\partial T)_\theta(T/\mathrm{K}) + (\mathrm{K}^2/12R)(\partial^2\Delta_r C_p^\infty/\partial T^2)_\theta(T/\mathrm{K})^2$

Table 5.6.2. Ives and Pryor equation (rearranged)—chemical equilibria

From Table 5.6.1,

$$\ln K(T) = \ln K(\theta/\mathrm{K}=1) + (K/2R)(\partial\Delta_r C_p^\infty/\partial T)_\theta(T/\mathrm{K}) + (\mathrm{K}^2/12R)(\partial^2\Delta_r C_p^\infty/\partial T^2)_\theta(T/\mathrm{K})^2$$

or $\ln K(T) = a_1 + a_2(T/\mathrm{K}) + a_3(T/\mathrm{K})^2$

$$d\ln K/dT = (a_2/K) + 2a_3(T/\mathrm{K}^2) = \Delta_r H^\infty(T)/RT^2$$

or $\Delta_r H^\infty(T)/R = a_2 T^2/\mathrm{K} + 2a_3 T^3/\mathrm{K}^2$
and $\Delta_r C_p^\infty(T)/R = (2a_2 T/\mathrm{K}) + (6a_3 T^2/\mathrm{K}^2)$

$$(1/R)(\partial\Delta_r C_p^\infty/\partial T) = 2a_2/\mathrm{K} + 12a_3 T/\mathrm{K}^2$$

$$(1/R)(\partial^2\Delta_r C_p^\infty/\partial T^2) = 12a_3/\mathrm{K} = (12/R)(\partial^2\Delta_r C_p^\infty/\partial T^2)_\theta$$

Table 5.6.3. Ives and Pryor equation

At ambient pressure: chloroethanoic acid (aq).
Reference [1]; Temperature range: $288.15 \leqslant \mathrm{T/K} \leqslant 308.15$
$a_1 = -14.52 \pm 2.17$; $a_2 = (5.946 \pm 1.45) \times 10^{-2}$; $a_3 = -(11.038 \pm 2.44) \times 10^{-5}$
At 298.15 K, $\Delta_r H^\infty(\mathrm{aq};)/\mathrm{kJ\,mol^{-1}} = -4.698$; $\Delta_r C_p^\infty(\mathrm{aq})/\mathrm{J\,K^{-1}\,mol^{-1}} = -194$.

References to section 5.6
[1] D. J. G. Ives; J. H. Pryor. *J. Chem. Soc.*, 1955, 2104.

5.7 FEATES AND IVES EQUATION

Some of the characteristic features of the Ives and Pryor equation (section 5.6) stem from the assumption that the second differential of $\Delta_r C_p^\infty$ with respect to temperature is constant. If this assumption is unacceptable we must choose a form for its dependence on temperature. Here we consider an approach (Table 5.7.1) which asserts that $(\partial^2 \Delta_r C_p^\infty / \partial T^2)$ is inversely proportional to temperature. It turns out that further assumptions have to be introduced if we require an equation having the form of the empirical equation proposed by Feates and Ives [1] (Table 5.7.2). Comparison shows that the parameters a_2 and a_3 are not independent but are related through the A parameter in Table 5.7.1. The immediate consequences of the new product term involving two functions of T are seen in the equation for the standard isobaric heat

Table 5.7.1. Feates and Ives equation—derivation

Fixed pressure. Reference temp/K $= \theta$
Assumption: $\partial^2 \Delta_r C_p^\infty / \partial T^2 = AR/T\mathrm{K}$
where A is independent of temperature.

$$\int_\theta^T \mathrm{d}\{\partial \Delta_r C_p^\infty / \partial T\} = \int_\theta^T \{AR/T\mathrm{K}\}\mathrm{d}T$$

Then, $(\partial \Delta_r C_p^\infty / \partial T)_T = (\partial \Delta_r C_p^\infty / \partial T)_\theta + \{AR/\mathrm{K}\} \ln (T/\mathrm{K}) - \{AR/\mathrm{K}\} \ln (\theta/\mathrm{K})$
Assumption: $(\partial \Delta_r C_p^\infty / \partial T)_\theta = (AR/\mathrm{K}) \ln (\theta/\mathrm{K})$

Then, $\int_\theta^T \mathrm{d}\Delta_r C_p^\infty = (AR/\mathrm{K}) \int_\theta^T \ln (T/\mathrm{K})\mathrm{d}T$

$$\Delta_r C_p^\infty(T) = \Delta_r C_p^\infty(\theta) + (AR/\mathrm{K})[T \ln (T/\mathrm{K}) - T]_\theta^T$$
$$= \Delta_r C_p^\infty(\theta) + (AR/\mathrm{K})[T \ln (T/\mathrm{K}) - \theta \ln (\theta/\mathrm{K}) - T + \theta]$$

Assumption: $\Delta_r C_p^\infty(\theta) = (AR/\mathrm{K})[\theta \ln (\theta/\mathrm{K}) - \theta]$
Then, $\Delta_r C_p^\infty(T) = (AR/\mathrm{K})[T \ln (T/\mathrm{K}) - T]$

Hence*, $\Delta_r H^\infty(T) = \Delta_r H^\infty(\theta) + (AR/\mathrm{K}) \int_\theta^T [T \ln (T/\mathrm{K}) - T]\mathrm{d}T$
Then

$$\Delta_r H^\infty(T) = \Delta_r H^\infty(\theta) + (AR/\mathrm{K})[(T^2/2) \ln (T/\mathrm{K}) - (T^2/4) - (T^2/2)]_\theta^T$$

or $\Delta_r H^\infty(T) = \Delta_r H^\infty(\theta) + (AR/\mathrm{K})[(T^2/2) \ln (T/\mathrm{K}) - (\theta^2/2) \ln (\theta/\mathrm{K}) - (3T^2/4) + (3\theta^2/4)]$
Assumption: $\Delta_r H^\infty(\theta) = (AR/\mathrm{K})[(\theta^2/2) \ln (\theta/\mathrm{K}) - (3\theta^2/4)]$
Hence $\Delta_r H^\infty(T) = (AR/\mathrm{K})[(T^2/2) \ln (T/\mathrm{K}) - (3T^2/4)]$

Check: $\mathrm{d}[(T^2/2) \ln T - T^2/4 - T^2/2] = T \ln T + (T^2/2)(1/T) - T/2 - T = T \ln T - T$

Table 5.7.2. Feates and Ives equation

From Table 5.7.1 and the van't Hoff equation,

$$\ln K(T) = \ln K(\theta) + (A/\mathrm{K})\int_{\theta}^{T}[(1/2)\ln(T/\mathrm{K}) - (3/4)]\mathrm{d}T$$

$$= \ln K(\theta) + (A/\mathrm{K})[(1/2)T\ln(T/\mathrm{K}) - (T/2) - (3T/4)]_{\theta}^{T}$$

Hence, $\ln K(T) = \ln K(\theta) + (A/\mathrm{K})[(T/2)\ln(T/\mathrm{K}) - (\theta/2)\ln(\theta/\mathrm{K}) - (5/4)(T - \theta)]$
or $\ln K(T) = \{\ln K(\theta) + (A/\mathrm{K})[-(\theta/2)\ln(\theta/\mathrm{K}) + (5\theta/4]\} + (A/2\mathrm{K})T\ln(T/\mathrm{K}) -$
$(5A/4\mathrm{K})T$

Table 5.7.3. Feates and Ives equation (conventional); chemical equilibria

At fixed pressure
From Table 5.7.2. $\ln K(T) = a_1 + a_2 T/\mathrm{K} + a_3(T/\mathrm{K})\ln(T/\mathrm{K})$
where $a_1 = \{\ln K(\theta) + (A/\mathrm{K})[-(\theta/2)\ln(\theta/\mathrm{K}) + (5\theta/4]\}$

$$a_2 = (A/2\mathrm{K}) \text{ and } a_3 = -(5A/4\mathrm{K})$$

$$\Delta_r H^{\infty}(T) = (R/\mathrm{K})[a_2 T^2 + a_3 T^2 + a_3 T^2 \ln(T/\mathrm{K})]$$

$$\Delta_r C_p^{\infty}(T) = (R/\mathrm{K})[2a_2 T + 3a_3 T + 2a_3 T\ln(T/\mathrm{K})]$$

$$\mathrm{d}\Delta_r C_p^{\infty}/\mathrm{d}T = (R/\mathrm{K})[2a_2 + 5a_3 + 2a_3\ln(T/\mathrm{K})]$$

$$\mathrm{d}^2\Delta_r C_p^{\infty}/\mathrm{d}T^2 = 2a_3 R/(T\mathrm{K}) \text{ [see Table 5.7.1]}$$

Limit$(T \rightarrow 0)\ln K$ is indeterminate. Limit$(T \rightarrow \infty)\ln K$ is indeterminate
At $T/\mathrm{K} = 1$, $\ln K = a_1 + a_2$

capacities of reaction. The second and subsequent derivatives of $\Delta_r C_p^{\infty}$ with respect to temperature are determined by the A parameter and must alternate in sign. In fact Feates and Ives expressed concern that the form of the equation prejudices the sign and magnitude of derived thermodynamic parameters. This concern is particularly strong where estimates are sought of parameters describing the dependence of $\Delta_r C_p^{\infty}(T)$ on temperature. Application of the Feates and Ives equation (Table 5.7.3) to experimental data is illustrated (Table 5.7.4) using the dependence on temperature of acid dissociation constants for 2-cyanoethanoic acid.

Table 5.7.4. 2-Cyanoethanoic acid(aq) (At fixed pressure)

Reference [1]: Temperature range. $278.15 \leqslant T/\mathrm{K}$ 318.15
$K(\mathrm{aq}; 2981.5\,\mathrm{K}; \mathrm{expt}) = 3.3876 \times 10^{-3}$
$a_1 = -21.062 \pm 0.446$, $a_2 = 0.373 \pm 0.010$, $a_3 = -(5.646 \pm 0.15) \times 10^{-2}$,
st. error $= 1.104 \times 10^{-3}$
At 298.15 K, $\Delta_r H^{\infty}(\mathrm{aq};)/\mathrm{kJ\,mol^{-1}} = -3.62$; $\Delta_r C_p^{\infty}(\mathrm{aq})/\mathrm{J\,K^{-1}\,mol^{-1}} = -164$.

References to section 5.7
[1] F. S. Feates; D. J. Ives. *J. Chem. Soc.*, 1956, 2798.

5.8 HARNED AND EMBREE EQUATION

Harned and Embree [1] described the dependence on temperature of the dissociation constant for methanoic acid in aqueous solution using an equation based on a linear dependence of $\Delta_r C_p^\infty$ on temperature. Consequently the enthalpy of reaction $\Delta_r H^\infty$ has

Table 5.8.1. Harned and Embree Equation

At fixed pressure.
Reference temperature /K $= \theta$ Assumption; $(\partial \Delta_r C_p^\infty / \partial T) =$ constant
Then, $\Delta_r C_p^\infty(T) = \Delta_r C_p^\infty(\theta) + (\partial \Delta_r C_p^\infty / \partial T)_\theta (T - \theta)$
or $\Delta_r C_p^\infty(T) = \Delta_r C_p^\infty(\theta) - \theta(\partial \Delta_r C_p^\infty / \partial T)_\theta + T(\partial \Delta_r C_p^\infty / \partial T)_\theta$
If $BR = \Delta_r C_p^\infty(\theta) - \theta(\partial \Delta_r C_p^\infty / \partial T)_\theta$,
then, $\Delta_r C_p^\infty(T) = BR + T(\partial \Delta_r C_p^\infty / \partial T)_\theta$

Hence, $\Delta_r H^\infty(T) = \Delta_r H^\infty(\theta) + \int_\theta^T [BR + T(\partial \Delta_r C_p^\infty / \partial T)_\theta] \mathrm{d}T$

$$= \Delta_r H^\infty(\theta) + [BRT + (\partial \Delta_r C_p^\infty / \partial T)_\theta (T^2/2)]_\theta^T$$

Hence,

$$\Delta_r H^\infty(T) = \Delta_r H^\infty(\theta) + BR(T - \theta) + (\partial \Delta_r C_p^\infty / \partial T)_\theta [(T^2/2) - (\theta^2/2)]$$

If $RK\beta = \Delta_r H^\infty(\theta) - RB\theta - (\partial \Delta_r C_p^\infty / \partial T)_\theta (\theta^2/2)$
then, $\Delta_r H^\infty(T) = (RK\beta) + (RBT) + (1/2)(\partial \Delta_r C_p^\infty / \partial T)_\theta T^2$

Hence, $\ln K(T) = \ln K(\theta) + (1/R)\int_\theta^T [(RK\beta/T^2) + (RB/T) + (1/2)(\partial \Delta_r C_p^\infty / \partial T)_\theta] \mathrm{d}T$

or, $\ln K(T) = \ln K(\theta) + (1/R)[-(RK\beta/T) + RB \ln T + (1/2)(\partial \Delta_r C_p^\infty / \partial T)_\theta T]_\theta^T$
$\ln K(T) = \ln K(\theta) + \{K\beta[(1/\theta) - (1/T)] + B \ln(T/\theta) + (1/2R)(\partial \Delta_r C_p^\infty / \partial T)_\theta (T - \theta)\}$

Table 5.8.2. Harned and Embree equation—conventional form

From Table 5.8.1,

$$\ln K(T) = \{\ln K(\theta) + [K\beta/\theta] - B \ln(\theta/K) - (\theta/2R)(\partial \Delta_r C_p^\infty / \partial T)_\theta\}$$

$$- (K\beta/T) + B \ln(T/K) + (1/2R)(\partial \Delta_r C_p^\infty / \partial T)_\theta T$$

Then, $\ln K(T) = a_1 + a_2(K/T) + a_3 \ln(T/K) + a_4 T/K$
where $a_1 = \{\ln K(\theta) + [K\beta/\theta] - B \ln(\theta/K) - (\theta/2R)(\partial \Delta_r C_p^\infty / \partial T)_\theta\}$; $a_2 = -\beta$; $a_3 = B$;
$a_4 = (1/2R)(\partial \Delta_r C_p^\infty / \partial T)_\theta$

Table 5.8.3. Harned and Embree equation—derived parameters

From Table 5.8.1, at fixed pressure,

$$\ln K(T) = a_1 - (\mathrm{K}\beta/T) + B\ln(T/\mathrm{K}) + (1/2R)(\partial\Delta_r C_p^\infty/\partial T)_\theta T$$

$$\partial \ln K/\partial T = (\mathrm{K}\beta/T^2) + B\mathrm{K}/T + (1/2R)(\partial\Delta_r C_p^\infty/\partial T)_\theta$$

$$\Delta_r H^\infty(T)/R = \mathrm{K}\beta + B\mathrm{K}T + (1/2R)(\partial\Delta_r C_p^\infty/\partial T)_\theta T^2$$

$$\Delta_r C_p^\infty(T)/R = B\mathrm{K} + (T/R)(\partial\Delta_r C_p^\infty/\partial T)_\theta$$

and $(\partial\Delta_r C_p^\infty/\partial T) = (\partial\Delta_r C_p^\infty/\partial T)_\theta$; assumption (Table 5.8.1)

Table 5.8.4. Extrema in ln *K*

At extremum in $\ln K(T)$, $\Delta_r H^\infty(T) = 0$
Hence, $(\partial\Delta_r C_p^\infty/\partial T)_\theta\{(T^2/2) - (\theta^2/2)\} + RB(T-\theta) + \Delta_r H^\infty(\theta) = 0$

Table 5.8.5. *o*-Fluorobenzoic acid(aq)

Ambient pressure.
Reference [2]. Temperature range: $278.15 \leqslant T/\mathrm{K} \leqslant 373.15$

$$K(\text{aq: } 298.15\,\mathrm{K}; \text{expt}) = 3.399 \times 10^{-4}$$

$a_1 = 256.83 \pm 32.05$, $a_2 = -(84.89 \pm 8.93) \times 10^2$, $a_3 = -43.87 \pm 5.55$,
st. error $= 2.268 \times 10^{-3}$
At 298.15 K, $\Delta_r H^\infty(\text{aq})/\mathrm{kJ\,mol^{-1}} = -4.45$; $\Delta_r C_p^\infty(\text{aq})/\mathrm{J\,K^{-1}\,mol^{-1}} = -138.5$

Table 5.8.6. Dependence of Gibbs energy of reaction on temperature

From Table 5.8.1, assuming pressure $p \simeq p^0$.

$$\Delta_r G^0(\theta) = -R\theta\ln K(\theta) \qquad \Delta_r G^0(T) = -RT\ln K(T)$$

$$\ln K(\theta) = -\Delta_r G^0(\theta)/R\theta \qquad \ln K(T) = -\Delta_r G^0(T)/RT$$

Then $-\Delta_r G^0(T) = -RT[\{\Delta_r G^0(\theta)/R\theta\} + \{\mathrm{K}\beta[(1/\theta) - (1/T)]\} + B\ln(T/\theta)$
$+ (\partial\Delta_r C_p^\infty/\partial T)_\theta(T-\theta)/2R]$

or $\Delta_r G^0(T) = (R\mathrm{K}\beta) + T\{[\Delta_r G^0(\theta)/\theta] - [R\mathrm{K}\beta/\theta] + RB\ln(\theta/\mathrm{K})$
$+ [(\theta/2)(\partial\Delta_r C_p^\infty/\partial T)_\theta]\} - RBT\ln(T/\mathrm{K}) - (\partial\Delta_r C_p^\infty/\partial T)_\theta T^2$

or $\Delta_r G^0(T)/R\mathrm{K} = a_1 + a_2 T/\mathrm{K} + a_3(T/\mathrm{K})\ln(T/\mathrm{K}) + a_4(T/\mathrm{K})^2$

a quadratic dependence on temperature. We derive in Table 5.8.1 an equation for the corresponding dependence of $\ln K$ on T about reference temperature θ. The terms in this equation can be grouped (Table 5.8.2) to produce the original equation which involves the variables T^{-1}, $\ln(T/\mathrm{K})$ and T. The four derived parameters are complicated functions of the thermodynamic parameters at temperature θ. Some simplification emerges if θ is set at 1.0 K (Table 5.8.2). Integration is straightforward between two limits, T and 1.0 K, at which the equilibrium constants are $K(T)$ and $K(T/\mathrm{K} = 1.0)$. A least squares analysis yields estimates of four variables, a_i for $i = 1$–4 (Tables 5.8.3 and 5.8.4). The four-term equation (Table 5.8.3) describes the dependences on temperature of the dissociation constants [2] for *o*-fluorobenzoic acid(aq) (Table 5.8.5). A related approach fits the dependence on temperature of $\Delta_r G^0$ using the equation in Table 5.8.6 where the temperature variables are T, $\ln(T/\mathrm{K})$ and $(T/\mathrm{K})^2$.

References to section 5.8

[1] H. S. Harned; N. D. Embree. *J. Amer. Chem. Soc.*, 1934, **56**, 1042.

[2] L. E. Strong; C. Van Waes; K. H. Doolittle. *J. Soln. Chem.*, 1982, **11**, 237.

5.9 MAGEE, RI AND EYRING EQUATION

In this and the next few sections, the analysis takes rather a different route. The treatments described above start out with equations describing the dependence of heat capacity terms on temperature. The emphasis now switches to understanding the dependence of thermodynamic parameters on temperature about a reference tempera-

---

Table 5.9.1. Magee, Ri and Eyring equation—chemical equilibria

At fixed pressure $p \simeq p^0$

At temperature T, $\ln K(T) = -[\Delta_r H^\infty(T)/RT] + [\Delta_r S^0(T)/R]$

At temperature θ, $\ln K(\theta) = -[\Delta_r H^\infty(\theta)/R\theta] + [\Delta_r S^0(\theta)/R]$

Then, $\ln \mathrm{K}(T) = \ln K(\theta) - \Delta_r H^\infty(T)/RT + \Delta_r H^\infty(\theta)/R\theta + \Delta_r S^0(T)/R - \Delta_r S^0(\theta)/R$

Taylor expansions:

(a) for the enthalpy using three terms;

$$\Delta_r C_p^\infty(T) = \Delta_r H^\infty(\theta) + \Delta_r C_p^\infty(\theta)(T-\theta) + (1/2)\{\partial\Delta_r C_p^\infty/\partial T\}_\theta(T-\theta)^2$$

(b) for the isobaric heat capacity term using two terms;

$$\Delta_r C_p^\infty(T) = \Delta_r C_p^\infty(\theta) + \{\partial\Delta_r C_p^\infty/\partial T\}_\theta(T-\theta)$$

Consider the binomial expansion of $(1/T)$ as follows,

$$(1/T) = [1/\{T-\theta) + \theta\}] = (1/\theta)[1 + \{(T-\theta)/\theta\}]^{-1}$$
$$= (1/\theta)[1 - \{(T-\theta)/\theta\} + \{(T-\theta)^2/\theta^2\} - \{(T-\theta)^3/\theta^3\} + \cdots]$$

We confine attention to the

(1) first two terms in the Taylor expansion for $\Delta_r C_p^\infty(T)$, and

(2) first three terms in the binomial expansion of $(1/T)$

Table 5.9.2. Entropy term

From $\Delta_r S^0(T) = \Delta_r S^0(\theta) + \int_\theta^T [\Delta_r C_p^\infty / T] dT$
Using the Taylor expansion, we express $\Delta_r C_p^\infty(T)$ in terms of $\Delta_r C_p^\infty(\theta)$ and $(\partial \Delta_r C_p^\infty / \partial T)_\theta$

$$\Delta_r S^0(T) = \Delta_r S^0(\theta) + \int_\theta^T [(1/T)[\Delta_r C_p^\infty(\theta) + (\partial \Delta_r C_p^\infty / \partial T)_\theta (T - \theta)]] dT$$

We incorporate the Taylor expansion for T^{-1}

$$\Delta_r S^0(T) = \Delta_r S^0(\theta) + (1/\theta) \int_\theta^T I dT$$

where $I = [\Delta_r C_p^\infty(\theta) - \Delta_r C_p^\infty(\theta)\{(T-\theta)/\theta\} + \Delta_r C_p^\infty(\theta)\{(T-\theta)^2/\theta^2\} \ldots]$
$+ [(\partial \Delta_r C_p^\infty / \partial T)_\theta (T-\theta) - (\partial \Delta_r C_p^\infty / \partial T)_\theta ((T-\theta)^2/\theta) \ldots]$

All terms higher than those shown are ignored.

Hence, $\Delta_r S^0(T) = \Delta_r S^0(\theta) + (1/\theta)[\Delta_r C_p^\infty(\theta) T - \Delta_r C_p^\infty(\theta)\{(T-\theta)^2/2\theta\}]_\theta^T$
$+ (1/\theta)[\Delta_r C_p^\infty(\theta)\{T-\theta)^3/3\theta^2\} + (\partial \Delta_r C_p^\infty / \partial T)_\theta ((T-\theta)^2/2)]_\theta^T$
$- (1/\theta)[(\partial \Delta_r C_p^\infty / \partial T)_\theta ((T-\theta)^3/3\theta)]_\theta^T$

Hence, $\Delta_r S^0(T) = \Delta_r S^0(\theta) + [\Delta_r C_p^\infty(\theta)(T-\theta)/\theta] - [\Delta_r C_p^\infty(\theta)(T-\theta)^2/2\theta^2]$
$+ [\Delta_r C_p^\infty(\theta)(T-\theta)^3/3\theta^3] + [(\partial \Delta_r C_p^\infty / \partial T)_\theta (T-\theta)^2/2\theta]$
$- [(\partial \Delta_r C_p^\infty / \partial T)_\theta (T-\theta)^3/3\theta^2]$

Hence, (by grouping terms in powers of $T - \theta$),

$$\begin{aligned}\Delta_r S^0(T) &= \Delta_r S^0(\theta) + [\Delta_r C_p^\infty(\theta)/\theta](T-\theta) \\ &- \{[\Delta_r C_p^\infty(\theta)/2\theta^2] - [\partial \Delta_r C_p^\infty / \partial T]_\theta (1/2\theta)\}(T-\theta)^2 \\ &+ \{[\Delta_r C_p^\infty(\theta)/3\theta^3] - [(\partial \Delta_r C_p^\infty / \partial T)_\theta (1/3\theta^2)]\}(T-\theta)^3\end{aligned}$$

ture θ. Here the latter is usually near the median of the measured temperature range. A seminal treatment (though rarely cited) is described by Eyring and coworkers [1].

Taylor expansions express $\Delta_r H^\infty(T)$ at temperature T in terms of $\Delta_r H^\infty(\theta)$, and $\Delta_r C_p^\infty(T)$ in terms of $\Delta_r C_p^\infty(\theta)$ (Table 5.9.1). We restrict the analysis to cases where the isobaric heat capacity of reaction $\Delta_r C_p^\infty(T)$ is a linear function of $T - \theta$. An expansion (Table 5.9.2) relates entropies of reaction at temperatures T and θ, the binomial theorem providing an equation for T^{-1} in terms of $T - \theta$. $\ln K(T)$ is expressed as a function of T in terms of thermodynamic parameters for reaction at temperature θ (Table 5.9.3). The final equation (Tables 5.9.4 and 5.9.5) is a polynomial in $T - \theta$. A given thermodynamic quantity (e.g. $\Delta_r C_p^\infty(\theta)$) occurs in more than one term in the polynomial (Table 5.9.5). For example, the product of $(T - \theta)^3$ in the polynomial is

Table 5.9.3. Dependence of $K(T)$ on temperature

From Tables 5.9.1 and 5.9.2. Substitute for $\Delta_r H^\infty(T)$ and $\Delta_r S^0(T)$,

$$\begin{aligned}\ln \mathrm{K}(T) = {} & \ln \mathrm{K}(\theta) - (1/RT)\{\Delta_r H^\infty(\theta) + \Delta_r C_p^\infty(\theta)(T-\theta) \\ & + (1/2)(\partial\Delta_r C_p^\infty/\partial T)_\theta (T-\theta)^2\} + (\Delta_r H^\infty(\theta)/R\theta) \\ & + (\Delta_r C_p^\infty(\theta)/\theta R)(T-\theta) \\ & - \{[\Delta_r C_p^\infty(\theta)/2\theta^2 R] - [(\partial\Delta_r C_p^\infty/\partial T)_\theta(1/2\theta R)\}(T-\theta)^2 \\ & + \{[\Delta_r C_p^\infty(\theta)/3\theta^3 R] - [(\partial\Delta_r C_p^\infty/\partial T)_\theta(1/3R\theta^2)]\}(T-\theta)^3\end{aligned}$$

We use the binomial expansion for T^{-1} (Table 5.9.1) and ignore all terms in $(T-\theta)^4$ and in higher powers of $T-\theta$,

$$\begin{aligned}\ln K(T) = {} & \ln K(\theta) - \{\Delta_r H^\infty(\theta)/R\}\{(1/\theta) - [(T-\theta)/\theta^2] + [(T-\theta)^2/\theta^3] \\ & - [(T-\theta)^3/\theta^4]\} + [\Delta_r H^\infty(\theta)/R\theta] \\ & - \{\Delta_r C_p^\infty(\theta)/R\}(T-\theta)\{(1/\theta) - [(T-\theta)/\theta^2] + [(T-\theta)^3/\theta^3]\} \\ & - [\partial\Delta_r C_p^\infty/\partial T]_\theta(1/2R)(T-\theta)^2\{(1/\theta) - [(T-\theta)/\theta^2]\} \\ & + \{\Delta_r C_p^\infty(\theta)/\theta R\}(T-\theta) - \{[\Delta_r C_p^\infty(\theta)/2\theta^2 R] \\ & - [(\partial\Delta_r C_p^\infty/\partial T)_\theta(1/2\theta R)]\}(T-\theta)^2 \\ & + \{[\Delta_r C_p^\infty(\theta)/3\theta^3 R] - [(\partial\Delta_r C_p^\infty/\partial T)_\theta(1/3R\theta^2)]\}(T-\theta)^3\end{aligned}$$

Table 5.9.4. Magee, Ri and Eyring equation

From Table 5.9.3 we collect terms in ascending powers of $T-\theta$

$$\begin{aligned}\ln K(T) = {} & \ln K(\theta) + \{\Delta_r H^\infty(\theta)/R\theta^2\}(T-\theta) \\ & + \{[-\Delta_r H^\infty(\theta)/R\theta^3] + [\Delta_r C_p^\infty(\theta)/2R\theta^2]\}(T-\theta)^2 \\ & + \{[\Delta_r H^\infty(\theta)/R\theta^4] - [(2/3)(\Delta_r C_p^\infty(\theta)/R\theta^3] \\ & + [(1/6)(\partial\Delta_r C_p^\infty/\partial T)_\theta(1/R\theta^2)]\}(T-\theta)^3\end{aligned}$$

---

related to $\Delta_r H^\infty(\theta)$ and $\{\partial\Delta_r C_p^\infty/\partial T\}_\theta$. The derived thermodynamic parameters are calculated from the a_i parameters (Tables 5.9.5 and 5.9.6). An example of the analysis is reported in Table 5.9.7 (see also Fig. 5.9.1).

In the event that $\ln K(T)$ is a maximum at temperature θ, the second term in the modified equation (Table 5.9.8) yields $\Delta_r C_p^\infty(\theta)$. If $b_2 = 0$, the Harned and Embree quadratic equation (Table 5.9.1) is recovered.

The equation (Table 5.9.4) has considerable merit from a computational point of view in that a linear least squares method is used. The major problem emerges from unravelling the thermodynamic parameters.

Table 5.9.5. Dependence of $K(T)$ on temperature—polynomial

From Table 5.9.4 (at fixed pressure):

$$\ln K(T) = a_1 + a_2(T-\theta) + a_3(T-\theta)^2 + a_4(T-\theta)^3$$

$$\mathrm{d}\ln K/\mathrm{d}T = a_2 + a_3 2(T-\theta) + a_4 3(T-\theta)^2$$

$$\Delta_r H^\infty(T)/R = a_2 T^2 + a_3 2T^2(T-\theta) + a_4 3T^2(T-\theta)^2$$

$$\Delta_r C_p^\infty(T)/R = 2a_2 T + 4a_3 T(T-\theta) + 2a_3 T^2 + 6a_4 T(T-\theta)^2 + 6a_4 T^2(T-\theta)$$

$$(1/R)(\mathrm{d}\Delta_r C_p^\infty/\mathrm{d}T) = 2a_2 + 4a_3(T-\theta) + 8a_3 T + 6a_4(T-\theta)^2 + 24a_4 T(T-\theta) + 6a_4 T^2$$

Table 5.9.6. Dependence of $K(T)$ on temperature—polynomial

From Table 5.9.5 (at fixed pressure):

$$\ln K(T) = a_1 + a_2(T-\theta) + a_3(T-\theta)^2 + a_4(T-\theta)^3$$

where $a_1 = \ln K(\theta)$

$$a_2 = \Delta_r H^\infty(\theta)/R\theta^2$$

$$a_3 = \{[-\Delta_r H^\infty(\theta)/R\theta^3] + [\Delta_r C_p^\infty(\theta)/2R\theta^2]\}$$

$$= \{[-a_1/\theta] + [\Delta_r C_p^\infty(\theta)/2R\theta^2]\}$$

or $\Delta_r C_p^\infty(\theta) = 2[a_3 + (a_2/\theta)]R\theta^2$

Also

$$a_4 = [\Delta_r H^\infty(\theta)/R\theta^4] - (2/3)[\Delta_r C_p^\infty(\theta)/R\theta^3] + (1/6)[\partial\Delta_r C_p^\infty/\partial T]_\theta(1/R\theta^2)$$

$$= [a_2/\theta^2] - (4/3\theta)[a_3 + (a_2/\theta)] + (1/6)[\partial\Delta_r C_p^\infty/\partial T]_\theta(1/R\theta^2)$$

$$= -[a_2/3\theta^2] - [4a_3/3\theta] + (1/6)[\partial\Delta_r C_p^\infty/\partial T]_\theta(1/R\theta^2)$$

Table 5.9.7. 2,3,4,5,6-Pentafluorobenzoic acid(aq)

Reference [2]; ambient pressure.
Temperature range: $273.15 \leqslant T/K \leqslant 373.15$

$$K(\text{aq}; 298.15\,\text{K}; \text{expt}) = 3.295 \times 10^{-2}$$

Reference temperature $\theta/\text{K} = 318.15$; st. error $= 2.832 \times 10^{-3}$
$a_1 = -3.719324 \pm 0.000949$; $a_2 = (1.5759 \pm 0.0050) \times 10^{-2}$;
$a_3 = -(1.486 \pm 0.08702) \times 10^{-5}$; $a_4 = (1.89 \pm 0.287) \times 10^{-7}$
At 298.15 K, $\Delta_r H^\infty(\text{aq})/\text{kJ mol}^{-1} = -11.64$; $\Delta_r C_p^\infty(\text{aq})/\text{J K}^{-1}\,\text{mol}^{-1} = -113$.

Table 5.9.8. Dependence of $K(T)$ on temperature about $K(T; \max)$ at $T = \theta$

From Table 5.9.4, if $K(T)$ is a maximum at $T = \theta$, $\Delta_r H^{\infty}(\theta) = 0$

$$\ln K(T) = b_1 + b_2(T - \theta)^2 + b_3(T - \theta)^3$$

where $b_1 = \ln K(\theta)$; $b_2 = \Delta_r C_p^{\infty}(\theta)/2R\theta^2$
and $b_3 = -[2\Delta_r C_p^{\infty}(\theta)/3R\theta^3] + (1/6R\theta^2)[\partial\Delta_r C_p^{\infty}/\partial T]_{\theta}$

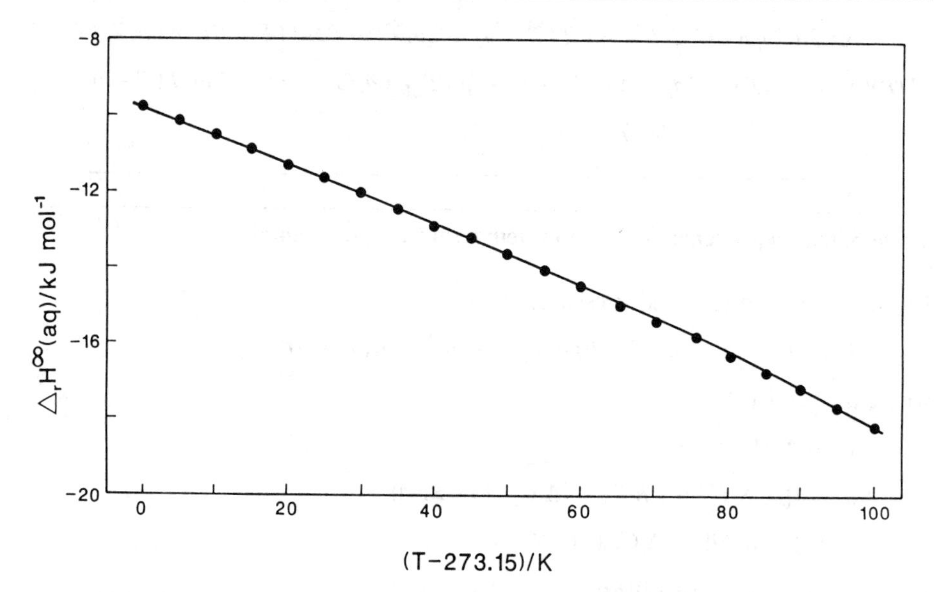

Fig. 5.9.1. Dependence on temperature of the limiting enthalpy of dissociation of 2,3,4,5,6-pentafluorobenzoic acid(aq); (see Table 5.9.7).

References to section 5.9

[1] J. L. Magee; T. Ri; H. Eyring. *J. Chem. Phys.*, 1941, **9**, 419.
[2] L. E. Strong, C. L. Brummel; P. Lindower. *J. Soln. Chem.*, 1987, **16**, 105.

5.10 CLARKE AND GLEW EQUATION

Chemical equilibria

The Clarke–Glew equation [1] has justifiably attracted considerable attention. Taylor expansions are used to describe the dependence of $\Delta_r H^{\infty}(T)$ at temperature T about $\Delta_r H^{\infty}(\theta)$ at θ, and for $\Delta_r C_p^{\infty}(T)$ about $\Delta_r C_p^{\infty}(\theta)$ (Table 5.10.1). The equation for $\Delta_r C_p^{\infty}(T)$ is integrated with respect to $\ln(T/\mathrm{K})$ over the range θ to T, yielding $\Delta_r S^0(T)$ in terms of $\Delta_r S^0(\theta)$ (Table 5.10.2). Equations for $\Delta_r H^{\infty}(T)$ and $\Delta_r S^0(T)$ express the difference between $\ln K(T)$ and $\ln K(\theta)$ (Table 5.10.3). We collect (Table 5.10.4) terms with common thermodynamic parameters (e.g. $\Delta_r H^{\infty}(\theta)$) to produce the final equation (Table 5.10.5). This collection process produces an equation which differs slightly

Table 5.10.1. Clarke–Glew equation

At pressure $p \simeq p^0$

At temperature T, $\Delta_r G^0(T) = -RT \ln K(T) = \Delta_r H^\infty(T) - T\Delta_r S^0(T)$

At temperature θ, $\Delta_r G^0(\theta) = -R\theta \ln K(\theta) = \Delta_r H^\infty(\theta) - \theta\Delta_r S^0(\theta)$

where $\Delta_r S^0(T) = \Delta_r S^0(\theta) + \int_\theta^T [\Delta_r C_p^\infty / T] \mathrm{d}T$

Assumptions: two Taylor expansions.

$$\Delta_r H^\infty(T) = \Delta_r H^\infty(\theta) + \Delta_r C_p^\infty(\theta)(T-\theta) + (1/2)[\partial \Delta_r C_p^\infty/\partial T]_\theta (T-\theta)^2 + (1/6)[\partial^2 \Delta_r C_p^\infty/\partial T^2]_\theta (T-\theta)^3 + (1/24)[\partial^3 \Delta_r C_p^\infty/\partial T^3]_\theta (T-\theta)^4$$

and, $$\Delta_r C_p^\infty(T) = \Delta_r C_p^\infty(\theta) + [\partial \Delta_r C_p^\infty/\partial T]_\theta (T-\theta) + (1/2)[\partial^2 \Delta_r C_p^\infty/\partial T^2]_\theta (T-\theta)^2 + (1/6)[\partial^3 \Delta_r C_p^\infty/\partial T^3]_\theta (T-\theta)^3$$

Table 5.10.2. Calculation of $\Delta_r S^0(T)$

From Table 5.10.1,

$$\Delta_r S^0(T) = \Delta_r S^0(\theta) + \int_\theta^T I \mathrm{d}T$$

$$I = [\Delta_r C_p^\infty(\theta)/T] + [\partial \Delta_r C_p^\infty/\partial T]_\theta [1 - (\theta/T)] + (1/2)[\partial^2 \Delta_r C_p^\infty/\partial T^2]_\theta [T - (2\theta) + (\theta^2/T)] + (1/6)[\partial^3 \Delta_r C_p^\infty/\partial T^3]_\theta [T^2 - (3T\theta) + (3\theta^2) - (\theta^3/T)]$$

$$\Delta_r S^0(T) = \Delta_r S^0(\theta) + \big[[\Delta_r C_p^\infty(\theta) \ln(T/\mathrm{K})] + [\partial \Delta_r C_p^\infty/\partial T]_\theta [T - (\theta \ln(T/\mathrm{K}))]\big]_\theta^T + \big[(1/2)[\partial^2 \Delta_r C_p^\infty/\partial T^2]_\theta [(T^2/2) - (2\theta T) + (\theta^2 \ln(T/\mathrm{K}))]\big]_\theta^T + (1/6)[\partial^3 \Delta_r C_p^\infty/\partial T^3]_\theta \big[(T^3/3) - (3T^2\theta/2) + (3\theta^2 T) - (\theta^3 \ln(T/\mathrm{K}))\big]_\theta^T$$

Hence,

$$\Delta_r S^0(T) = \Delta_r S^0(\theta) + \Delta_r C_p^\infty(\theta) \ln(T/\theta) + [\partial \Delta_r C_p^\infty/\partial T]_\theta [T - \theta - (\theta \ln(T/\theta))] + (1/2)[\partial^2 \Delta_r C_p^\infty/\partial T^2]_\theta [(T^2/2) - (\theta^2/2) - (2\theta T) + (2\theta^2) + (\theta^2 \ln(T/\theta))] + (1/6)[\partial^3 \Delta_r C_p^\infty/\partial T^3]_\theta [(T^3/3) - (\theta^3/3) - (3T^2\theta/2) + (3\theta^3/2) + (3\theta^2 T) - (3\theta^3) - (\theta^3 \ln(T/\theta))]$$

from that described in Table 5.9.4, although the starting points are similar. The quantities $f(T;\theta)$ are complicated and the relationships between the various $f(T;\theta)$ terms are not obvious. Overall the equation has the form of a series and so one must judge using statistical criteria where in a given application the series must be terminated. For many purposes it is unlikely that the data warrant calculation of terms beyond $\Delta_r C_p^\infty(\theta)$, i.e. a three-term Clarke–Glew Equation (Table 5.10.5).

Table 5.10.3. Development of the Clarke–Glew equation

From Table 5.10.1

$$-R\ln K(T) = [\Delta_r H^\infty(T)/T] - \Delta_r S^0(T)$$

and $$-R\ln K(\theta) = [\Delta_r H^\infty(\theta)/\theta] - \Delta_r S^0(\theta)$$

Therefore, $$-R\ln K(T) = -R\ln K(\theta) + \{[\Delta_r H^\infty(T)/T] - [\Delta_r H^\infty(\theta)/\theta]\} - [\Delta_r S^0(T) - \Delta_r S^0(\theta)]$$

Hence, from Tables 5.10.1 and 5.10.2:

$$\begin{aligned}-R\ln K(T) = {} & -R\ln K(\theta) + \{\Delta_r H^\infty(\theta)T\} + \{\Delta_r C_p^\infty(\theta)[(T-\theta)/T]\} \\ & + \{(1/2)[\partial\Delta_r C_p^\infty/\partial T]_\theta[T-(2\theta)+(\theta^2/T)]\} \\ & + \{(1/6)[\partial^2\Delta_r C_p^\infty/\partial T^2]_\theta[(T-\theta)^3/T]\} + \{(1/24)[\partial^3\Delta_r C_p^\infty/\partial T^3]_\theta \\ & \times [(T-\theta)^4/T]\} - \{\Delta_r H^\infty(\theta)/\theta\} - \{\Delta_r C_p^\infty(\theta)\ln(T/\theta)\} \\ & - \{[\partial\Delta_r C_p^\infty/\partial T]_\theta[T-\theta-(\theta\ln(T/\theta))]\} - \{(1/2)[\partial^2\Delta_r C_p^\infty/\partial T^2]_\theta \\ & \times [(T^2/2)+(3\theta^2/2)-(2\theta T)+(\theta^2\ln(T/\theta))]\} \\ & - \{(1/6)[\partial^3\Delta_r C_p^\infty/\partial T^3]_\theta[(T^3/3)-(11\theta^3/6)-(3T^2\theta/2) \\ & + (3\theta^2 T)-(\theta^2\ln(T/\theta))]\}\end{aligned}$$

Table 5.10.4. Collecting terms

From Table 5.10.3,

$$\begin{aligned}R\ln K(T) = {} & R\ln K(\theta) - \{[\Delta_r H^\infty(\theta)/T] - [\Delta_r H^\infty(\theta)/\theta]\} \\ & + \Delta_r C_p^\infty(\theta)\{-[(T-\theta)/T] + \ln(T/\theta)\} + [\partial\Delta_r C_p^\infty/\partial T]_\theta \\ & \times \{(T/2)+\theta-(\theta^2/2T)+T-\theta-[\theta\ln(T/\theta)]\} + [\partial^2\Delta_r C_p^\infty/\partial T^2]_\theta \\ & \times \{-(T^2/6)+(3T\theta/6)-(3\theta^2/6)+(\theta^3/6T)+(T^2/4) \\ & + (3\theta^2/4)-(\theta T)+[(\theta^2/2)\ln(T/\theta)\} + [\partial^3\Delta_r C_p^\infty/\partial T^3]_\theta \\ & \times \{-(T^3/24)+(4T^2\theta/24)-(6T\theta^2/24)+(4\theta^3/24)-(\theta^4/24T) \\ & + (T^3/18)-(11\theta^3/36)-(3T^2\theta/12)+(\theta^2 T/2)-[(\theta^3/6)\ln(T/\theta)]\}\end{aligned}$$

An example of the application of the Clarke–Glew equation is reported in Table 5.10.6 with reference to the dependence on temperature of the self-dissociation constant for water at ambient pressure (see also Fig. 5.10.1). We commented in connection with the Valentiner equation (Table 5.4.4 and 5.4.5) on the correlation

Table 5.10.5. Clarke and Glew equation; chemical equilibria

From Table 5.10.4, simplifying each product term of a thermodynamic property.

Equation: $\ln K(T) = \ln K(\theta) + \{\Delta_r H^{\infty}(\theta)/R\}\{(1/\theta) - (1/T)\} + \{\Delta_r C_p^{\infty}(\theta)/R\}$
$\times \{(\theta/T) - 1 + \ln(T/\theta)\} + \{\theta/2R\}[\partial \Delta_r C_p^{\infty}/\partial T]_\theta$
$\times \{(T/\theta) - (\theta/T) - [2\ln(T/\theta)]\} + \{\theta^2/12R\}[\partial^2 \Delta_r C_p^{\infty}/\partial T^2]_\theta$
$\times \{(T^2/\theta^2) - (6T/\theta) + 3 + (2\theta/T) + [6\ln(T/\theta)]\}$
$+ \{\theta^3/72R\}[\partial^3 \Delta_r C_p^{\infty}/\partial T^3)]_\theta\{(T^3/\theta^3) - (6T^2/\theta^2) + (18T/\theta)$
$- 10 - (3\theta/T) - [12\ln(T/\theta)]\}$

Table 5.10.6. Self-dissociation of water

Reference [2]. Ambient pressure
Temperature range; $273.15 \leqslant T/\mathrm{K}\ 333.15$

$pK_w(298.15\,\mathrm{K};\ \mathrm{expt}) = 14.004$

Reference temperature $\theta/\mathrm{K} = 298.15$

$a_1 = -37.941 \pm 3.9 \times 10^{-4}$, $a_2 = (53.696 \pm 0.0025) \times 10^4$,

$a_3 = -224.92 \pm 1.53$, $a_4 = 1.623 \pm 0.23$,

At 298.15 K, $\Delta_r H^{\infty}(\mathrm{aq})/\mathrm{kJ\,mol^{-1}} = 53.696$

$\Delta_r C_p^{\infty}(\mathrm{aq})/\mathrm{J\,K^{-1}\,mol^{-1}} = -224.9$

$[\mathrm{d}\Delta_r C_p^{\infty}(\mathrm{aq})/\mathrm{d}T]/\mathrm{J\,K^{-1}\,mol^{-1}} = 1.62$

Table 5.10.7. Correlation matrix

From Table 5.10.6

	1	2	3	4
1	1.0000			
2	−0.204	1.0000		
3	−0.737	0.398	1.0000	
4	0.29	−0.898	−0.585	1.0000

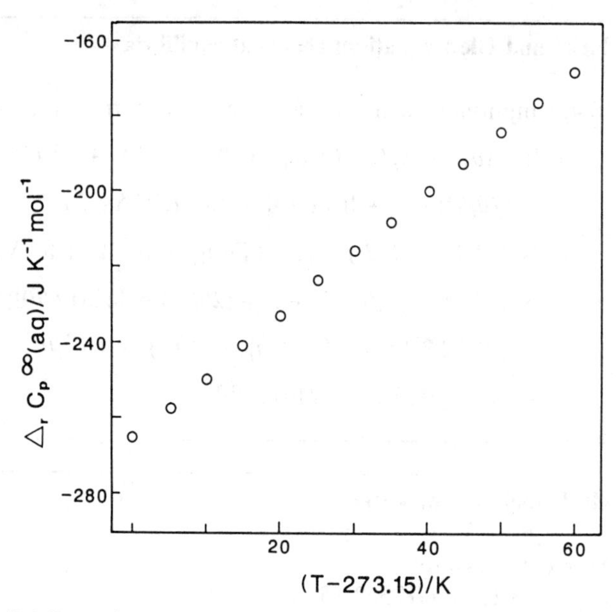

Fig. 5.10.1. Dependence on temperature of the limiting isobaric heat capacity of reaction, $\Delta_r C_p^{\infty}$(aq) for the self-dissociation of water (see Table 5.10.6).

Table 5.10.8. Clarke–Glew equation; chemical kinetics

From Tables 5.10.5

$$\ln(k^{\$}\mathrm{K}/T) = \ln(k^{\$}(\theta)\mathrm{K}/\theta) + \{\Delta^{\neq}H^{\infty}(\theta)/R\}\{(1/\theta)-(1/T)\} + \{\Delta^{\neq}C_p^{\infty}(\theta)/R\} \times \{(\theta/T) - 1 + \ln(T/\theta)\} + \{\theta/2R\}[\partial\Delta^{\neq}C_p^{\infty}/\partial T]_{\theta}\{(T/\theta)-(\theta/T) - [2\ln(T/\theta)]\} + \{\theta^2/12R\}[\partial^2\Delta^{\neq}C_p^{\infty}/\partial T^2]_{\theta}\{(T^2/\theta^2)-(6T/\theta)+3 + (2\theta/T) + [6\ln(T/\theta)]\} + \{\theta^3/72R\}[\partial^3\Delta^{\neq}C_p^{\infty}/\partial T^3]_{\theta}\{(T^3/\theta^3) - (6T^2/\theta^2) + (18T/\theta) - 10 - (3\theta/T) - [12\ln(T/\theta)]\}$$

matrix, pointing out that the diagonal elements are very close to unity. We contrast that state of affairs with the correlation matrix associated with the Clarke–Glew equation (Table 5.10.7). The off-diagonal elements are not unity, indicating that the temperature functions in the Clarke–Glew equation are not closely related.

Chemical kinetics

The Clarke–Glew equation can be written in a form which expresses the dependence of rate constants on temperature about a reference temperature, θ (Table 5.10.8).

The algebra associated with the temperature functions in the Clarke–Glew method looks a little formidable. Nevertheless there are advantages in this procedure.

Chemists have a strong hunch at the start of a numerical analysis concerning the reliability of estimates of $\Delta_r C_p^\infty(T)$ and related temperature derivatives. There is therefore merit in a method which produces estimates of these parameters directly from the fitting procedures. If the Clarke–Glew Equation is written as a routine for a computer, the analysis routines are repeated, adding one more term in the equation, and terminated when the statistical significance of the next derived thermodynamic parameter is unacceptable.

References to section 5.10

[1] E. C. W. Clarke; D. N. Glew. *Trans. Faraday Soc.*, 1966, **62**, 539.
[2] C. P. Bezbourah; M. Filomena; G. F. C. Camoes; A. K. Covington; J. V. Dobson, *J. Chem. Soc. Faraday Tans. 1*, 1973, **69**, 949.

5.11 POLYNOMIAL EQUATION IN $(T - \theta)$

The procedures discussed in the previous two sections prompt [1, 2] the equation described in Table 5.11.1 which adopts from the outset a polynomial in $T - \theta$ to

Table 5.11.1. Polynomial equation: chemical equilibria

Equation: $\ln K(T) = \Sigma(j = 1; j = \mathrm{i}) a_j[(T - \theta)/\mathrm{K}]^{j-1}$

Table 5.11.2. Application; four-term equation

From Table 5.11.1

$$\ln K(T) = a_1 + a_2(T - \theta)/K + a_3[(T - \theta)/\mathrm{K}]^2 + a_4[(T - \theta)/\mathrm{K}]^3$$

$$\mathrm{limit}\,(T \to \theta) \ln K(T) = a_1$$

At temperature T, $\mathrm{d}\ln K/\mathrm{d}T = a_2/\mathrm{K} + 2a_3(T - \theta)/\mathrm{K}^2 + 3a_4(T - \theta)^2/\mathrm{K}^3$

$$\Delta_r H^\infty/R = a_2 T^2/\mathrm{K} + 2a_3 T^2(T - \theta)/\mathrm{K}^2 + 3a_4 T^2(T - \theta)^2/\mathrm{K}^3$$

$$\mathrm{limit}\,(T \to \theta) \quad \Delta_r H^\infty(T) = \Delta_r H^\infty(\theta) = R a_2 \theta^2/\mathrm{K}$$

$$\Delta_r C_p^\infty/R = 2a_2 T/\mathrm{K} + a_3\{2T^2 + 4T(T - \theta)\}/\mathrm{K}^2$$
$$+ a_4\{6a_4 T(T - \theta)^2 + 6a_4 T^2(T - \theta)\}/\mathrm{K}^3$$

$$\mathrm{limit}\,(T \to \theta) \quad \Delta_r C_p^\infty = \Delta_r C_p^\infty(\theta) = R[2a_2\theta/\mathrm{K} + 2\theta^2 a_3/\mathrm{K}^2]$$

$$(1/R)\{\mathrm{d}\Delta_r C_p^\infty/\mathrm{d}T\} = 2a_2/\mathrm{K} + a_3\{8T + 4(T - \theta)\}/\mathrm{K}^2$$
$$+ a_4\{6(T - \theta)^2 + 24T(T - \theta) + 6T^2\}/\mathrm{K}^3$$

$$\mathrm{limit}\,(T \to \theta)\ \{\mathrm{d}\Delta_r C_p^\infty/\mathrm{d}T\} = R[2a_2/\mathrm{K} + 8\theta a_3/\mathrm{K}^2 + 6a_4\theta^2/\mathrm{K}^3]$$

Table 5.11.3. Dependence of $K(T)$ on T about $K(\theta;\max)$ at θ

If $K(T)$ is a maximum at $T = \theta$, $\Delta_r H^\infty(\theta) = 0$, and $a_2 = 0$
From Table 5.11.2 equation:

$$\ln K(T) = \ln K(\theta;\max) + a_3(T-\theta)^2/\mathrm{K}^2 + a_4(T-\theta)^3/K^3$$

$$\Delta_r C_p^\infty(\theta) = 2R\theta^2 a_3/\mathrm{K}^2$$

or $a_3 = \Delta_r C_p^\infty(\theta)\mathrm{K}^2/2R\theta^2$

$$\{\partial\Delta_r C_p^\infty/\partial T\}_\theta = 8R\theta a_3/\mathrm{K}^2 + 6Ra_4\theta^2/\mathrm{K}^3$$

$$= 4\Delta_r C_p^\infty(\theta)/\theta + 6Ra_4\theta^2/\mathrm{K}^3$$

or $a_4 = \{[\mathrm{K}^3/6R\theta^2][\partial\Delta_r C_p^\infty/\partial T]_\theta\} - \{2\mathrm{K}^3\Delta_r C_p^\infty(\theta)/3R\theta^3\}$

Hence, $\ln K(T) = \ln K(\theta) + \{\Delta_r C_p^\infty(\theta)\mathrm{K}^2/2R\theta^2\}(T-\theta)^2/\mathrm{K}^2$

$$+ \{[\mathrm{K}^3/6R\theta^2][\partial\Delta_r C_p^\infty/\partial T]_\theta - [2\mathrm{K}^3\Delta_r C_p^\infty(\theta)/3\theta^3 R]\}(T-\theta)^3/\mathrm{K}^3$$

Table 5.11.4. Calculation of temperature where $K(T)$ is a maximum

From Table 5.11.2: if $\Delta_r H^\infty(T) =$ zero at T_m

$$a_2 + 2a_3(T_m - \theta)/\mathrm{K} + 3a_4(T_m - \theta)^2/\mathrm{K}^2 = 0$$

A quadratic in $(T_m - \theta)/\mathrm{K}$

Table 5.11.5. Polynomial in $(T-\theta)$; kinetic analysis

Equation: $\ln k^\$(T) = \Sigma(j=1; j=\mathrm{i})a_j\{(T-\theta)/\mathrm{K}\}^{(j-1)}$

---

express the dependence of $\ln K(T)$ on temperature about $\ln K(\theta)$. The consequences are explored in Table 5.11.2 of using a four-term polynomial.

If θ is the temperature at which $\ln K(T)$ is a maximum, the procedures can be rewritten to produce the equation described in Table 5.11.3. This method requires that the temperature θ is known. A better procedure uses the polynomial in Table 5.11.1 to determine T_m at $\{\ln K(T)\}_{\max}$, knowing that $\Delta_r H^\infty(T_m)$ is zero (Table 5.11.4).

A similar analysis can be undertaken in conjunction with kinetic data in which the dependence of $\ln\{k^\$(T)\mathrm{K}/T\}$ on temperature is fitted about $\ln\{k^\$(\theta)\mathrm{K}/\theta\}$ at temperature θ using a polynomial in $T-\theta$ (Tables 5.11.5 and 5.11.6).

Table 5.11.6. Application of polynomial equation; five terms

$$\ln\{k^{\$}(T)\mathrm{K}/T\} = \ln\{k^{\$}(\theta)\mathrm{K}/\theta\} + a_2[(T-\theta)/\mathrm{K}] + a_3[(T-\theta)/\mathrm{K}]^2 + a_4[(T-\theta)/\mathrm{K}]^3 + a_5[(T-\theta)/\mathrm{K}]^4$$

$$\mathrm{d}\ln\{k^{\$}\mathrm{K}/T\}/\mathrm{d}T = a_2/\mathrm{K} + 2a_3(T-\theta)/\mathrm{K}^2 + 3a_4(T-\theta)^2/\mathrm{K}^3 + 4a_5(T-\theta)^3/\mathrm{K}^4$$

$$\Delta^{\neq} H^{\infty}(T)/R = a_2 T^2/\mathrm{K} + 2a_3 T^2(T-\theta)/\mathrm{K}^2 + 3a_4 T^2(T-\theta)^2/\mathrm{K}^3 + 4a_5 T^2(T-\theta)^3/\mathrm{K}^4$$

At $T = \theta$, $\Delta^{\neq} H^{\infty}(\theta) = Ra_2\theta^2/\mathrm{K}$

$$\Delta^{\neq} C_{\mathrm{p}}^{\infty}(T)/R = 2a_2 T/\mathrm{K} + a_3\{2T^2 + 4T(T-\theta)\}/\mathrm{K}^2 + a_4\{6T(T-\theta)^2 + 6T^2(T-\theta)\}/\mathrm{K}^3 + a_5\{8T(T-\theta)^3 + 12T^2(T-\theta)^2\}/\mathrm{K}^4$$

or $\Delta^{\neq} C_{\mathrm{p}}^{\infty}(\theta)/R = 2a_2\theta/\mathrm{K} + 2a_3\theta^2/\mathrm{K}^2$

References to section 5.11

[1] M. J. Blandamer; J. M. W. Scott; R. E. Robertson. *J. Chem. Soc. Perkin II*, 1981, 447.

[2] M. J. Blandamer; J. Burgess; P. P. Duce; R. E. Robertson; J. M. W. Scott. *J. Chem. Soc. Perkin II*, 1982, 1157.

6

Dependence of equilibrium and rate constants on temperature—extended equations

6.1 INTRODUCTION

We can now draw together several strands in the argument presented. Overall the aim is to formulate equations which allow the dependence of $K(T)$ at pressure p on temperature to be analysed, leading to standard thermodynamic parameters for reaction. The argument proceeds as follows. $\ln K^0(T)$ is plotted against T^{-1}. For many systems, the plot is curved and the treatment described in Table 5.3.2 is inadequate. In many cases, however, the plot is close to linear. In fact this pattern seems so ingrained in chemists' thinking that curvature in the plot of $\ln K(T)$ against T^{-1} is often described as a 'deviation'. The term $a_1 \ln(T/\mathrm{K})$ in the Valentiner equation (Table 5.4.3) is said to take account of these deviations.

However there is no 'ideal' dependence and patterns formed by plots of $\ln K(T)$ against T^{-1} are characteristic of systems. The $a_3 \ln(T/\mathrm{K})$ term in Table 5.4.3 and the term $(\theta/T) - 1 + \ln(T/\theta)$ in the Clarke–Glew equation (Table 5.10.8) have the effect of bending the otherwise linear dependence. Hence the bend in the line is used to calculate in the first instance a heat capacity quantity. Faced with a table reporting $K(T)$ and T for a given system, we embark on the calculation starting, possibly, with the equation in Table 5.10.5. As new terms are added in the various equations, so we calculate a further temperature derivative of the enthalpy quantity, $\Delta_r H^\infty$. A new problem emerges. In this fitting procedure we have only one experimental variable, temperature. Unfortunately the terms in $f(T)$ in each equation are not always orthogonal [1]. In other words, they are inter-related. We noted that over the range $273 < T/\mathrm{K} < 333$, $\ln(T/\mathrm{K})$ is close to a linear function of $(T/\mathrm{K})^{-1}$ (see Valentiner Equation). $\ln(T/\mathrm{K})$ is also close to a linear function of T/K. Therefore derived parameters which are multipliers of $f(T)$ are linked and not as we might have hoped, independent.

There is a sense of conflict in the foregoing discussion between the tasks of (i) fitting data to an equation in an arithmetic exercise, and (ii) assigning physical significance to parameters calculated from the observed dependence of $K(T)$ on temperature. Nevertheless the procedures described in the previous chapter are closely related, having three common features: (i) integration of equations describing the dependence of $\Delta_r C_p^\infty$ on temperature, (ii) analysis of dependences about reference temperatures θ, and (iii) the linear least squares method of fitting the data to the equation for the dependence of $\ln K(T)$ on T. In this chapter we comment on procedures which differ from this common pattern.

References to section 6.1.1

[1] R. Gaboriaud, *Bull. Soc. Chem. France*, 1971, 1605.

6.2 ORTHOGONAL POLYNOMIALS

Many of the problems discussed in the previous section stem from the observation that the various temperature functions expressing the dependence of $\ln K(T)$ on temperature are not orthogonal. Consequently when new terms are added in the polynomial functional dependence in $T - \theta$, for example a term in $(T - \theta)^k$, the coefficients of $(T - \theta)^j$ for $j = 1$ to $j = k - 1$ change from those calculated using the polynomial terminated at the $(T - \theta)^{k-1}$ term.

A quantity z_j is defined in terms of T, the temperature interval, and the reference temperature θ (Table 6.2.1). The temperature functions in the equation for $K(T)$ are calculated as functions of z_j, these temperature functions satisfying the orthogonal condition [1–5]. The thermodynamic parameters are calculated as shown in Table 6.2.2. An example of the analysis is reported in Table 6.2.3.

Table 6.2.1. Orthogonal polynomials

At fixed pressure p.
Input: dependence of $\ln K(T)$ on T at equal temperature intervals.
Then, $\Delta = T(j + 1) - T(j)$ for the jth data point
Definition: If n is odd, $\theta = T[(n - 1)/2 + 1]$
If n is even, $\theta = \{T[(n/2)] + T[(n/2) + 1]\}/2$
At temperature T_j: by definition, $z_j = (T_j - \theta)/\Delta$
Then $\ln K = a_1 + a_2 P_1(z_j) + a_3 P(z_j) + a_4 P(z_j) + \dots$
where at temperature T_j

$$P_1(z_j) = z_j$$

$$P_2(z_j) = z_j^2 - ((n^2 - 1)/12.0)$$

$$P_3(z_j) = z_j^3 - ((3n^2 - 7)z_j/20.0)$$

In general terms,

$$P_{r+1} = P_r(z_j)P_1(z_j) - [r^2(n^2 - r^2)P_{r-1}(z_j)/\{4(4r^2 - 1)\}]$$

Table 6.2.2. Orthogonal polynomials

From Table 6.2.1, using a four-term polynomial,

$$\ln K(T) = a_1 + a_2[(T-\theta)/\Delta] + a_3\{[(T-\theta)/\Delta]^2 - [(n^2-1)/12]\} + a_4\{[(T-\theta)/\Delta]^3 - [(3n^2-7)/20][(T-\theta)/\Delta]\}$$

$$\mathrm{d}\ln K(T)/\mathrm{d}T = (a_2/\Delta) + a_3[2(T-\theta)/\Delta^2] + a_4\{[3(T-\theta)^2/\Delta^3] - [(3n^2-7)/(20\Delta)]\}$$

$$\Delta_r H^\infty(T)/R = [T^2 a_2/\Delta] + \{[2a_3/\Delta^2]T^2(T-\theta)\} + a_4\{[3T^2(T-\theta)^2/\Delta^3] - [T^2(3n^2-7)/(20\Delta)]\}$$

Hence, $\Delta_r H^\infty(\theta)/R = \{\theta^2 a_2/\Delta\} - \{a_4\theta^2(3n^2-7)/(20\Delta)\}$

$$\Delta_r C_p^\infty(T)/R = [2Ta_2/\Delta] + [2a_3/\Delta^2][2T(T-\theta) + T^2] + a_4\{[6T(T-\theta)^2/\Delta^3] + [6T^2(T-\theta)/\Delta^3] - [2T(3n^2-7)/(20\Delta)]\}$$

Hence, $\Delta_r C_p^\infty(\theta)/R = \{2\theta a_2/\Delta\} + \{2a_3\theta^2/\Delta^2\} - \{2\theta a_4(3n^2-7)/(20\Delta)\}$

Table 6.2.3. Di-isopropylcyanoethanoic acid(aq)

Data from reference [1]. Temperature $278.15 \leqslant T/\mathrm{K} \leqslant 318.15$

$\theta/\mathrm{K} = 298.15$

$K(\mathrm{aq}; \theta\,\mathrm{expt}) = 2.7816 \times 10^{-3}$

Analysis: $a_1 = -5.8879 \pm (1.73 \times 10^{-4})$

$a_2 = -(9.631 \pm 0.0128) \times 10^{-2}$

$a_3 = -(4.835 \pm 0.216) \times 10^{-4}$

$a_4 = (4.007 \pm 1.004) \times 10^{-5}$

$\Delta_r H^\infty(\mathrm{aq}; \theta)/\mathrm{kJ\,mol^{-1}} = -14.307$

$\Delta_r C_p^\infty(\mathrm{aq}; \theta)/\mathrm{J\,K^{-1}\,mol^{-1}} = -126.4$

References to section 6.2

[1] D. J. G. Ives; P. D. Marsden. *J. Chem. Soc.*, 1965, 2798.
[2] C. A. Bennett; N. L. Franklin. *Statistical Analysis in Chemistry and the Chemical Industry*, Wiley New York, 1954.
[3] D. J. G. Ives; P. G. N. Moseley. *J. Chem. Soc. B*, 1970, 1655.
[4] J. H. Ashby; E. M. Crook; S. P. Datta. *Biochem. J.* 1954, **56**, 190.
[5] N. W. Please. *Biochem. J.* 1954, **56**, 196.

6.3 HALLIWELL–STRONG EQUATION

In Chapter 5 we commented that in many cases comparison could be drawn between $\Delta_r H^\infty$ and $\Delta_r C_p^\infty$ estimated from the dependence of $K(T)$ on temperature and from calorimetric data. This separation and implied competition is perhaps undesirable. In fact there is merit in using precise calorimetric data for, say, $\Delta_r C_p^\infty$(sln; 298.2 K) in an analysis of the dependence of K(sln; T; p) on temperature. In this procedure we would start out with a calorimetric estimate of $\Delta_r C_p^\infty$(sln; θ; p) at temperature θ and select an equation [1] which describes the dependence of $\Delta_r C_p^\infty$(sln; T; p) on T. Two integrations with respect to T yield an equation for $\ln K(T)$.

In method A (Tables 6.3.1–6.3.4) $\Delta_r C_p^\infty(T)$ is expressed as a quadratic function of temperature. At each temperature T, the temperature variables α_i where $i = 1$ to 4, are calculated. A linear least squares analysis yields estimates of $\Delta_r S^0(\text{sln}; \theta)$, $\Delta_r H^\infty(\theta)$, $\Delta_r C_p^\infty(\theta)$, a_1 and a_2. Hence $\Delta_r H^\infty(T)$, $\Delta_r C_p^\infty(T)$ and $\Delta_r S^0(T; \text{sln})$ are calculated (Table 6.3.4).

The analysis described in Tables 6.3.5–6.3.8 is based on a slightly different equation for the dependence of $\Delta_r C_p^\infty(T)$ on temperature. Otherwise the algebra is similar to that outlined in Tables 6.3.1–6.3.4. Thus one term is linear in T whereas a third term shows a reciprocal dependence on T. The final equations in Tables 6.3.3 and 6.3.8 have a similar form. An application is summarized in Table 6.3.9.

The form of these equations does facilitate combination of data from various sources. In many cases $\Delta_r H^\infty(\theta)$ and $\Delta_r C_p^\infty(\theta)$ are known from calorimetric data where usually $\theta/\text{K} = 298.15$. If only $\Delta_r C_p^\infty(\theta)$ is known, then the analytical least

---

Table 6.3.1. Dependence of $\Delta_r C_p^\infty$ on temperature—method A

Assumption: $\Delta_r C_p^\infty / R = a_1 + a_2 T/\text{K} + a_3 (T/\text{K})^2$
Enthalpies:

$$\Delta_r H^\infty(T) - \Delta_r H^\infty(\theta) = \int_\infty^T \Delta_r C_p^\infty \, \mathrm{d}T$$

$\{\Delta_r H^\infty(T) - \Delta_r H^\infty(\theta)\}/R = a_1(T - \theta) + (a_2/2\text{K})(T^2 - \theta^2) + (a_3/3\text{K}^2)(T^3 - \theta^3)$
or $\Delta_r H^\infty(T) = h + R\{a_1 T + a_2(T^2/2\text{K}) + a_3(T^3/3\text{K}^2)\}$
where $h = \Delta_r H^\infty(\theta) - R\{a_1 \theta + a_2(\theta^2/2\text{K}) + a_3(\theta^3/3\text{K}^2)\}$
Entropies:
Assuming $p \simeq p^0$

$$\Delta_r S^0(T) - \Delta_r S^0(\theta) = \int_\theta^T \{\Delta_r C_p^\infty / T\} \, \mathrm{d}T$$

$$\{\Delta_r S^0(T) - \Delta_r S^0(\theta)\}/R = [a_1 \ln T + a_2 T/\text{K} + a_3 T^2/2\text{K}^2]_\theta^T$$

or $\Delta_r S^0(T) = s + R\{a_1 \ln(T/\text{K}) + a_2(T/\text{K}) + a_3(T^2/2\text{K}^2)\}$
where $s = \Delta_r S^0(\theta) - R\{a_1 \ln(\theta/\text{K}) + a_2(\theta/\text{K}) + a_3(\theta^2/2\text{K}^2)\}$
and $a_1 = \Delta_r C_p^\infty(\theta)/R - a_2(\theta/\text{K}) - a_3(\theta/\text{K})^2$

Table 6.3.2. Dependence of $K(T)$ on temperature

At temperature T, $R \ln K(T) = \Delta_r S^0(T) - \Delta_r H^\infty(T)/T$
Hence, $R \ln K(T) = s + R\{a_1 \ln(T/\mathrm{K}) + a_2(T/\mathrm{K}) + a_3(T^2/2\mathrm{K}^2)\}$
$- (1/T)[h + R\{a_1 T + a_2(T^2/2\mathrm{K}) + a_3(T^3/3\mathrm{K}^2)\}]$
or $R \ln K(T) = s + Ra_1\{\ln(T/\mathrm{K}) - 1\} - h/T + a_2 R(T/2\mathrm{K}) + a_3 R(T^2/6\mathrm{K}^2)$
Substitute for s and h;

$$R \ln K(T) = \Delta_r S^0(\theta) - a_1 R \ln(\theta/\mathrm{K}) - a_2 R(\theta/\mathrm{K}) - a_3 R(\theta^2/2\mathrm{K}^2)$$
$$+ a_1 R\{\ln(T/\mathrm{K}) - 1\} - [\Delta_r H^\infty(\theta) - a_1 R\theta - a_2 R(\theta^2/2\mathrm{K})$$
$$- a_3 R(\theta^3/3\mathrm{K}^2)]/T + a_2 R(T/2\mathrm{K}) + a_3 R(T^2/6\theta^2)$$

$$R \ln K(T) = \Delta_r S^0(\theta) - \Delta_r H^\infty(\theta)\{1/T\}$$
$$+ a_1 R\{-\ln(\theta/\mathrm{K}) + \ln(T/\mathrm{K}) - 1 + (\theta/T)\}$$
$$+ a_2 R\{-(\theta/\mathrm{K}) + (\theta^2/2T\mathrm{K}) + (T/2\mathrm{K})\}$$
$$+ a_3 R\{-(\theta^2/2\mathrm{K}^2) + (\theta^3/3T\mathrm{K}^2) + (T^2/6\mathrm{K}^2)\}$$

Substitute for a_1 in terms of $\Delta_r C_p^\infty(\theta)$;

$$R \ln K(T) = \Delta_r S^0(\theta) - \Delta_r H^\infty(\theta)\{1/T\}$$
$$+ \{\Delta_r C_p^\infty(\theta) - a_2 R(\theta/\mathrm{K}) - a_3 R(\theta^2/\mathrm{K}^2)\}\{\ln(T/\theta) - 1 + (\theta/T)\}$$
$$+ a_2 R\{-(\theta/\mathrm{K}) + (\theta^2/2T\mathrm{K}) + (T/2\mathrm{K})\}$$
$$+ a_3 R\{-(\theta^2/2\mathrm{K}^2) + (\theta^3/3T\mathrm{K}^2) + (T^2/6\,\mathrm{K}^2)\}$$

Grouping terms in a_2 and a_3:

$$R \ln K(T) = \Delta_r S^0(\theta) - \Delta_r H^\infty(\theta)\{1/T\} + \Delta_r C_p^\infty(\theta)\{\ln(T/\theta) - 1 + (\theta/T)\}$$
$$+ a_2 R\{-\theta \ln(T/\theta) + \theta - (\theta^2/T) - \theta + (\theta^2/2T) + (T/2)\}/\mathrm{K}$$
$$+ a_3 R\{-\theta \ln(T/\theta) + \theta^2 - (\theta^3/T) - (\theta^2/2) + (\theta^3/3T) + (T^2/6)\}/\mathrm{K}^2$$

Therefore

$$R \ln K(T) = \Delta_r S^0(\theta) - \Delta_r H^\infty(\theta)\{1/T\} + \Delta_r C_p^\infty(\theta)\{\ln(T/\theta) - 1 + (\theta/T)\}$$
$$+ a_2 R\{-\theta \ln(T/\theta) - (\theta^2/2T) + (T/2)\}/\mathrm{K}$$
$$+ a_3 R\{-\theta^2 \ln(T/\theta) + (\theta^2/2) - (2\theta^3/3T) + (T^2/6)\}/\mathrm{K}^2$$

---

squares method is based on the equation in Table 6.3.10, whereas if both $\Delta_r C_p^\infty(\theta)$ and $\Delta_r H^\infty(\theta)$ are known, the equation in Table 6.3.11 can be used.

Some indication of the underlying pattern to the dependence of $\Delta_r C_p^\infty(T)$ on temperature can be obtained using a double difference method. In the following we use the symbol Δ to indicate a difference, e.g. $\Delta T = T_2 - T_1$, $\Delta\alpha_1 = \alpha_1(T_2) - \alpha_1(T_1)$

Table 6.3.3. Halliwell–Strong equation—method A

From Table 6.3.2

$$\ln K(T) = \{\Delta_r S^0(\theta)/R\} - \{\Delta_r H^\infty(\theta)/RK\}\,\alpha_1 + \{\Delta_r C_p^\infty(\theta)/R\}\,\alpha_2 + a_2\alpha_3 + a_3\alpha_4$$

where

$$\alpha_1 = K/T$$

$$\alpha_2 = (\theta/T) - 1 + \ln(\theta/T)$$

$$\alpha_3 = (T/2K) - (\theta/K)\ln(T/\theta) - (\theta^2/2TK)$$

$$\alpha_4 = \{T^2/6K^2) + (\theta^2/2K^2) - (2\theta^3/3TK^2) - (\theta/K)^2\ln(T/\theta)$$

Table 6.3.4. Derived parameters at temperature T

Then from Tables 6.3.2 and 6.3.3, we obtain estimates using a linear least square analysis of:

(i) $\Delta_r S^0(\theta)$ (ii) $\Delta_r H^\infty(\theta)$
(iii) $\Delta_r C_p^\infty(\theta)$ (iv) a_2
(v) a_3

Then from Table 6.3.2.

$$a_1 = \Delta_r C_p^\infty(\theta)/R - a_2\theta/K - a_3(\theta/K)^2$$

Hence $\Delta_r C_p^\infty(T)/R = a_1 + a_2 T/K + a_3(T/K)^2$

$$\Delta_r H^\infty(T) = \Delta_r H^\infty(\theta) + R[a_1(T-\theta) + (a_2/2K)(T^2-\theta^2) + (a_3/3K^2)(T^3-\theta^3)]$$

Table 6.3.5. 3,5-Difluorobenzoic acid(aq)—Equation A

Data from [2]. Temperature $273.15 \leqslant T/K \leqslant 373.15$

$$K(\text{aq; expt; 298.16 K}) = 2.991 \times 10^{-4}$$

Analysis: $\theta/K = 318.15$

Equation A: $a_1 = -8.9867 \pm (2.23 \times 10^{-3})$
$a_2 = 272.33 \pm 0.710$
$a_3 = -18.981 \pm 0.0217$
$a_4 = (2.505 \pm 0.240) \times 10^{-3}$

$$\Delta_r H^\infty(\text{aq; 318.15 K})/\text{kJ mol}^{-1} = -2.26$$

$$\Delta_r S^0(\text{aq; 318.15 K})/\text{J K}^{-1}\text{mol}^{-1} = -74.71$$

$$\Delta_r C_p^\infty(\text{aq; 318.15 K})/\text{J K}^{-1}\text{mol}^{-1} = -157.$$

Table 6.3.6. Dependence of $K(T)$ on temperature—method B

Assuming $\Delta_r C_p^\infty(T)/R = a_1 + a_2(T/\mathrm{K}) + a_3(\mathrm{K}^2/T^2)$

$$\Delta_r H^\infty(T) = \Delta_r H^\infty(\theta) + \mathrm{R}\,[a_1 T + a_2(T^2/2\mathrm{K}) - a_3(\mathrm{K}^2/T)]_\theta^T$$

Then, $\Delta_r H^\infty(T) = \Delta_r H^\infty(\theta) + R[a_1(T-\theta) + (a_2/2\mathrm{K})(T^2-\theta^2) - a_3\mathrm{K}^2(T^{-1}-\theta^{-1})]$
$\Delta_r H^\infty(T) = h + (a_1 RT) + (a_2 RT^2/2\mathrm{K}) - (a_3 R\mathrm{K}^2/T)$
where $h = \Delta_r H^\infty(\theta) - (Ra_1\theta) - (a_2 R\theta^2/2\mathrm{K}) + (a_3 R\mathrm{K}^2/\theta)$

$$\Delta_r S^0(T) - \Delta_r S^0(\theta) = \int_\theta^T R\,[(a_1/T) + (a_2/\mathrm{K}) + (a_3\mathrm{K}^2/T^3)]\,\mathrm{d}T$$

$$= R[a_1 \ln(T/\mathrm{K}) + a_2(T/\mathrm{K}) - (a_3\mathrm{K}^2/2T^2)]_\theta^T$$

Then, $\Delta_r S^0(T) - \Delta_r S^0(\theta) = R[a_1\ln(T/\theta) + (a_2/\mathrm{K})(T-\theta) - (a_3\mathrm{K}^2/2)(T^{-2}-\theta^{-2})]$
or $\Delta_r S^0(T) = s + R[a_1\ln(T/\mathrm{K}) + a_2(T/\mathrm{K}) - (a_3\mathrm{K}^2/2T^2)]$
where $s = \Delta_r S^0(\theta) - a_1 R\ln(\theta/\mathrm{K}) - a_2 R(\theta/\mathrm{K}) + a_3 R\mathrm{K}^2/(2\theta^2)$

Table 6.3.7. Dependence of $K(T)$ on temperature—method B

Following the analysis given in Table 6.3.6.

$$R\ln K(T) = \Delta_r S^0(T) - \Delta_r H^\infty(T)/T$$
$$= \{s + a_1 R\ln(T/\mathrm{K}) + a_2 R(T/\mathrm{K}) - (a_3 R\mathrm{K}^2/2T^2)\}$$
$$- \{h + a_1 RT + (a_2 RT^2/2\mathrm{K}) - (a_3 R\mathrm{K}^2/T)\}/T$$

or $R\ln K(T) = s + a_1 R\{\ln(T/\mathrm{K}) - 1\} - h/T + (a_2 RT/2\mathrm{K}) + (a_3 R\mathrm{K}^2/2T^2)$

Hence, $R\ln K(T) = \Delta_r S^0(\theta) - R[a_1\ln(\theta/\mathrm{K}) + a_2(\theta/\mathrm{K}) - (a_3\mathrm{K}^2/2\theta^2) - a_1\{\ln(T/\mathrm{K}) - 1\}]$
$$- \Delta_r H^\infty(\theta)\{1/T\} + R[(a_1\theta/T) + a_2(\theta^2/2T\mathrm{K}) - (a_3 R\mathrm{K}^2/\theta T)$$
$$+ (a_2 T/2\mathrm{K}) + (a_3\mathrm{K}^2/2T^2)]$$

or $R\ln K(T) = \Delta_r S^0(\theta) - \Delta_r H^\infty(\theta)\{1/T\} + Ra_1[\ln(T/\theta) - 1 + (\theta/T)]$
$$+ a_2 R[-(\theta/\mathrm{K}) + (\theta^2/2T\mathrm{K}) + (T/2\mathrm{K})]$$
$$+ a_3 R[(\mathrm{K}^2/2\theta^2) + (\mathrm{K}^2/2T^2) - (\mathrm{K}^2/\theta T)]$$

Substituting for a_1 in terms of $\Delta_r C_p^\infty(\theta)$ (Table 6.3.6)

$$R\ln K(T) = \Delta_r S^0(\theta) - \Delta_r H^\infty(\theta)\{1/T\} + \{\Delta_r C_p^\infty(\theta) - a_2 R(\theta/\mathrm{K})$$
$$- (a_3 RK^2/\theta^2\}\{\ln(T/\theta) - 1 + (\theta/T)\} + a_2 R\{-(\theta/\mathrm{K}) + (\theta^2/2T\mathrm{K})$$
$$+ (T/2\mathrm{K})\} + a_3 R\{(\mathrm{K}^2/2\theta^2) + (\mathrm{K}^2/2T^2) - (\mathrm{K}^2/\theta T)\}$$

Therefore, $R\ln K(T) = \Delta_r S^0(\theta) - \Delta_r H^\infty(\theta)\{1/T\} + \Delta_r C_p^\infty(\theta)\{\ln(T/\theta) - 1 + (\theta/T)\}$
$$+ a_2 R\{-(\theta/\mathrm{K})\ln(T/\theta) - (\theta^2/2T\mathrm{K}) + (T/2\mathrm{K})\}$$
$$+ a_3 R\{-(\mathrm{K}^2/\theta^2)\ln(T/\theta) + (3\mathrm{K}^2/2\theta^2) - (2\mathrm{K}^2/T) + (\mathrm{K}^2/2T^2)\}$$

Table 6.3.8. Halliwell–Strong equation—method B

$$\ln K(T) = \Delta_r S^0(\theta)/R + \{\Delta_r H^\infty(\theta)/R\text{K}\}\,\beta_1 + \{\Delta_r C_p^\infty(\theta)/R\}\,\beta_2 + a_2\beta_3 + a_4\beta_4$$

where

$\beta_1 = -1/T$

$\beta_2 = \ln(T/\theta) - 1 + (\theta/T)$

$\beta_3 = (T/2\text{K}) - (\theta/\text{K}) - (\theta/\text{K})\ln(T/\theta) - (\theta^2/2T\text{K})$

$\beta_4 = (3\text{K}^2/2\theta^2) - (\text{K}^2/\theta^2)\ln(T/\theta) - (2\text{K}^2/\theta T) + (\text{K}^2/2T^2)$

A least squares analysis yields estimates of

(i) $\Delta_r S^0(\theta)$ (ii) $\Delta_r H^\infty(\theta)$ (iii) $\Delta_r C_p^\infty(\theta)$
(iv) a_2 (v) a_4

where $a_1 = [\Delta_r C_p^\infty(\theta)/R] - (a_2 T/\text{K}) - (a_3\text{K}^2/T^2)$

$$\Delta_r C_p^\infty(T)/R = a_1 + a_2(T/\text{K}) + a_3(\text{K}^2/T^2)$$

and $\Delta_r H^\infty(T) = \Delta_r H^\infty(\theta) + R\{a_1(T-\theta) + (a_2/2\text{K})(T^2-\theta^2) - (a_3\text{K}^2)(T^{-1}-\theta^{-1})\}$

Table 6.3.9. 2,5-Difluorobenzoic acid(aq)—Equation B

Data from [2]. Temperature $273.15 \leqslant T/\text{K} \leqslant 373.15$

$$K(\text{aq; expt; }298.15\text{ K}) = 7.805 \times 10^{-4}$$

Analysis: $\theta/\text{K} = 318.15$

Equation A: $a_1 = -9.886 \pm (2.34 \times 10^{-3})$
$a_2 = -823.98 \pm 0.74$
$a_3 = 15.399 \pm 0.023$
$a_4 = -(9.093 \pm 2.53) \times 10^{-3}$

$$\Delta_r H^\infty(\text{aq; }318.15\text{ K})/\text{kJ mol}^{-1} = -6.85$$

$$\Delta_r C_p^\infty(\text{aq; }318.15\text{ K})/\text{J K}^{-1}\text{mol}^{-1} = -128.$$

Table 6.3.10. Halliwell–Strong equation incorporating known $\Delta_r C_p^\infty(\theta)$

From Tables 6.3.3 or 6.3.7; Methods A and B.
By definition; $Y = \ln K(T) - [\Delta_r C_p^\infty(\theta)/R]\alpha_2$
Then $Y = \{\Delta_r S^0(\theta)/R\} - \{\Delta_r H^\infty(\theta)/R\}\,\alpha_1 + a_2\alpha_3 + a_3\alpha_4$

Table 6.3.11. Halliwell–Strong equation incorporating known $\Delta_r C_p^\infty(\theta)$ and $\Delta_r H^\infty(\theta)$

From Tables 6.3.3 or 6.3.7; Methods A and B.
By definition; $Y = \ln K(T) + [\Delta_r H^\infty(\theta)/R]\alpha_1 - [\Delta_r C_p^\infty(\theta)/R]\alpha_2$
Then $Y = \{\Delta_r S^0(\theta)/R\} + a_2\alpha_3 + a_3\alpha_4$

Table 6.3.12. Double difference analysis

From Tables 6.3.3 or 6.3.7; Methods A and B.

$$R\ln K(T) = \Delta_r S^0(\theta) - \{\Delta_r H^\infty(\theta)/\mathrm{K}\}\alpha_1 + \Delta_r C_p^\infty(\theta)\alpha_2 + a_2R\alpha_3 + a_3R\alpha_4$$

Then $R\Delta\ln K(T) = -\{\Delta_r H^\infty(\theta)/\mathrm{K}\}\Delta\alpha_1 + \Delta_r C_p^\infty(\theta)\Delta\alpha_2 + a_2R\alpha_3 + a_3R\Delta\alpha_4$

By definition; $Q = R\Delta\ln K(T)/\Delta\alpha_1$

$$Q = -\{\Delta_r H^\infty(\theta)/\mathrm{K}\} + \Delta_r C_p^\infty(\theta)\{\Delta\alpha_2/\Delta\alpha_1\} + a_2R\{\Delta\alpha_3/\Delta\alpha_1\} + a_3R\{\Delta\alpha_4/\Delta\alpha_1\}$$

Then, $\Delta Q = \Delta_r C_p^\infty(\theta)\Delta\{\Delta\alpha_2/\Delta\alpha_1\} + a_2R\Delta\{\Delta\alpha_3/\Delta\alpha_1\} + a_3R\Delta\{\Delta\alpha_4/\Delta\alpha_1\}$
By definition, $Z = \Delta Q/\Delta\{\Delta\alpha_2/\Delta\alpha_1\}$

$$\beta_1 = \Delta\{\Delta\alpha_3/\Delta\alpha_1\}/\Delta\{\Delta\alpha_2/\Delta\alpha_1\}$$

$$\beta_2 = \Delta\{\Delta\alpha_4/\Delta\alpha_1\}/\Delta\{\Delta\alpha_2/\Delta\alpha_1\}$$

Then, $Z = \Delta_r C_p^\infty(\theta) + a_2R\beta_1 + a_3R\beta_2$

and $\Delta R\ln K(T) = R\ln K(T_2) - R\ln K(T_1)$. Then the Halliwell–Strong equations describing the dependence of $K(T)$ on temperature can be compared at two temperatures T_2 and T_1. A double difference quantity is defined where the leading term is $\Delta_r C_p^\infty(\theta)$ (Table 6.3.12). The dependence of Z on β_1 is examined. If Z is independent of β_1, then a_2 and a_3 are likely to be zero. If Z is a linear function of β_1, then a_3 is zero. If Z is a quadratic function of β_1, then a_2 and a_3 are non-zero. The underlying calculation is a formidable exercise in arithmetic except in terms of a computer program. Temperature intervals in the calculation of differences should be restricted to 20, 30 and 40 K. Hence with 21 data points spanning, at equal intervals, the range $273.15 < T/\mathrm{K} < 373.15$, there are 13, 9 and 5 estimates of Z respectively.

References to section 6.3

[1] H. F. Halliwell; L. E. Strong. *J. Phys. Chem.*, 1985, **89**, 4137.
[2] L. E. Strong; C. L. Brummel; P. Lindower. *J. Soln. Chem.*, 1987, **16**, 105.

6.4 GURNEY EQUATION

The methods described above are not based on models for the chemical reaction described by $K(T)$. By way of contrast Gurney suggested [1] that for a weak acid in

Table 6.4.1. Gurney equation

$$-RT\ln K(T) = a_1 RK + a_2 RK \exp(T/\theta)$$

$$\ln K(T) = a_1(K/T) + a_2 \exp(T/\theta)$$

aqueous solution, $\Delta_r G^0(T)$ is determined by two contributions arising from (i) electrical interactions between solvent and solutes, described by a term $a_2 \exp(T/\theta)$ and (ii) non-electric interactions which are independent of temperature. The Gurney equation (Table 6.4.1) contains three unknowns, including the temperature θ.

Because $\ln K(T)$ has in part an exponential dependence on (T/θ), the task of fitting data to this equation requires a non-linear least squares procedure [2]. The dependence of $\exp(T/\theta)$ on T has a maximum at $T = \theta$. According to Gurney, θ is determined by the relative permittivity of the solvent and is independent of acid. There is a subtle difference between the analysis exemplified by the Gurney equation and those discussed earlier. There the dependence of $\ln K(T)$ on T is fitted to a general equation. In the Gurney treatment the data are fitted to a model.

References to section 6.4

[1] R. W. Gurney. *Ionic Processes in Solution*. McGraw-Hill, New York, 1953.
[2] M. J. Blandamer; J. Burgess; P. P. Duce; R. E. Robertson; J. W. M. Scott. *Can. J. Chem.* 1981, **59**, 2845.

6.5 SIGMA METHOD

Two particular features stand out from the analytical methods described so far. First most equations are based on a polynomial or series function with the requirement that statistical analysis determines where they should be terminated. Second, most equations describe the dependence of $\ln K(T)$ on T with equations which imply that K is an extremum at some temperature. The first feature means that subsequent differentiation with respect to temperature can amplify errors which mask important features in derived parameters. The second feature means that a plot of $\ln K(T)$ against T is markedly curved which is not necessarily a helpful property to account for in the curve-fitting process. Instead analyses are sought which use simple rectilinear plots.

The Sigma method [1] (see Table 6.5.1), developed from the procedure used by Everett and Wynne-Jones [2], is applied to a set of data reporting $\Delta_r G^0(T)$ at fixed temperature intervals, ΔT. The data are first examined to gain some idea of the underlying pattern. A table is constructed: the first column sets out the n-estimates of $\Delta_r G^0(T)$; the second column reports the $n-1$ differences ε_1 between $\Delta_r G^0(T)$ and $\Delta_r G^0(T + \Delta T)$. The next column lists the $n-2$ second differences, $\varepsilon_2, \ldots$. At the foot of each column the means of the differences are noted; i.e. $\Sigma(\varepsilon_1)/(n-1)$, $(\Sigma\varepsilon_2)/(n-2)$, $(\Sigma\varepsilon_3)/(n-3)\ldots$. In effect these differences are estimates of the first, second, third... differentials; i.e. $\Delta^j(\Delta_r G)/\Delta T^j$ for $j = 1, 2, 3, \ldots$. Inspection reveals a $(j+1)$ column

Table 6.5.1. Sigma method; thermodynamics

At fixed pressure.
At pressure $p \simeq p^0$; at temperature T, $\Delta_r G^0(T) = \Delta_r H^\infty(T) - T\Delta_r S^0(T)$

$$\Delta_r G^0(T) = \Delta_r H^\infty(\theta) + \left[\int_\theta^T \Delta_r C_p^\infty \, dT\right] - T\left[\Delta_r S^0(\theta) + \int_\theta^T \Delta_r C_p^\infty \, d\ln T\right]$$

Table 6.5.2. Cyanoethanoic acid(aq)

T/K	$\Delta_r G^0$/kJ mol^{-1}	ε_1	ε_2	ε_3	ε_4
278.15	13.01739				
		245.49			
283.15	13.26289		19.19		
		264.69		−7.43	
288.15	13.52758		11.76		13.80
		276.45		6.37	
293.15	13.80403		18.14		−14.95
		294.59		−8.57	
298.15	14.09861		9.56		13.39
		304.15		4.81	
303.15	14.40276		14.38		−7.56
		318.52		−2.74	
308.15	14.72129		11.64		7.19
		330.16		4.44	
313.15	15.05145		16.08		
		346.24			
318.15	15.39769				
		mean	mean	mean	mean
		297.54	14.39	−0.52	2.37
		±34.1	±3.6	±2.37	12.99

Table 6.5.3. Entropy of reaction

With $\Delta_r S^0(T + \Delta T/2) = [\Delta_r G^0(T + \Delta T) - \Delta_r G^0(T)]/\Delta T$
Then $\Delta_r S^0(T + \Delta T/2) = a_1 + \Delta_r C_p^\infty(\text{aq}) \ln(T/\text{K})$
For the present data set; $\Delta T/2 = 2.5$ K

$$a_1 = \Delta_r S^0(3.5\,\text{K}); \text{ i.e. } \theta = 3.5\,\text{K}$$

Table 6.5.4. Linear least squares analysis

Cyanoethanoic acid(aq); equation in Table 6.10.1

$a_1 = 884.98 \pm 22.71$ $a_2 = 165.79 \pm 3.99$

$\Delta_r S^0(3.5\,\mathrm{K})/\mathrm{J\,K^{-1}} = 884.98$

$\Delta_r C_p^\infty(\mathrm{aq})/\mathrm{J\,K^{-1}\,mol^{-1}} = -165.8$

Table 6.5.5. Enthalpy of reaction

At temperature T, $\Delta_r G^0(T) = \Delta_r H^\infty(T) - T\Delta_r S^0(T)$
But $\Delta_r H^0(T) = \Delta_r H^\infty(\theta) + (T-\theta)\Delta_r C_p^\infty$
and $\Delta_r S^0(T) = \Delta_r S^0(\theta) + \Delta_r C_p^\infty \ln(T/\theta)$
Then $\Delta_r G^0(T) = \Delta_r H^\infty(\theta) + (T-\theta)\Delta_r C_p^\infty - T[\Delta_r S^0(\theta) + \Delta_r C_p^\infty \ln(T/\theta)]$
or $\Delta_r G^0(T) + \theta\Delta_r C_p^\infty + \Delta_r C_p^\infty \ln(T/\theta)$

$$= \Delta_r H^\infty(\theta) - [\Delta_r S^0(\theta) - \Delta_r C_p^\infty]\,T$$

Hence a linear least squares analysis yields intercept, $\Delta_r H^\infty(\theta)$
where $\Delta_r H^0(T) = \Delta_r H^\infty(\theta) + (T-\theta)\,\Delta_r C_p^\infty$
for $\Delta_r G^0(T) + \theta\Delta_r C_p^\infty + \Delta_r C_p^\infty \ln(T/\theta) = a_1 + a_2 T$

Table 6.5.6. Enthalpy of reaction; linear least squares analysis

Cyanoethanoic acid(aq); equation—see Table 6.5.5

$a_1 = (45.156 \pm 0.011) \times 10^3$ $a_2 = -1050.7 \pm 3.8 \times 10^{-2}$

$\Delta_r H^\infty(\theta)/\mathrm{kJ\,mol^{-1}} = 45.156$

$\Delta_r H^\infty(298.15\,\mathrm{K})/\mathrm{kJ\,mol^{-1}} = -3.69$

in which the differences alternate in sign and where the mean of ε_{j+1} differences is larger than the mean of ε_j differences.

In effect the difference $\delta\{\Delta_r G^0(T)\}$ at fixed temperature intervals δT become, with sign reversed, an indication of the dependence on temperature of the entropy of reaction. Thus $-\delta\{\Delta_r G^0(T)\}/\delta T$ is equivalent to $\Delta_r S^0(\mathrm{s\,ln};T)$. The difference ε_2 is a measure of the dependence of $\Delta_r S^0$ on temperature $[\mathrm{d}\Delta_r S^0/\mathrm{d}\ln T = \Delta_r C_p^\infty]$ and hence the limiting heat capacity of reaction. The difference ε_3 is a measure of the dependence on temperature of $\Delta_r C_p^\infty$. The analysis is described by reference to a particular example. We use the data discussed by Ives and Mosely [1] for the acid dissociation

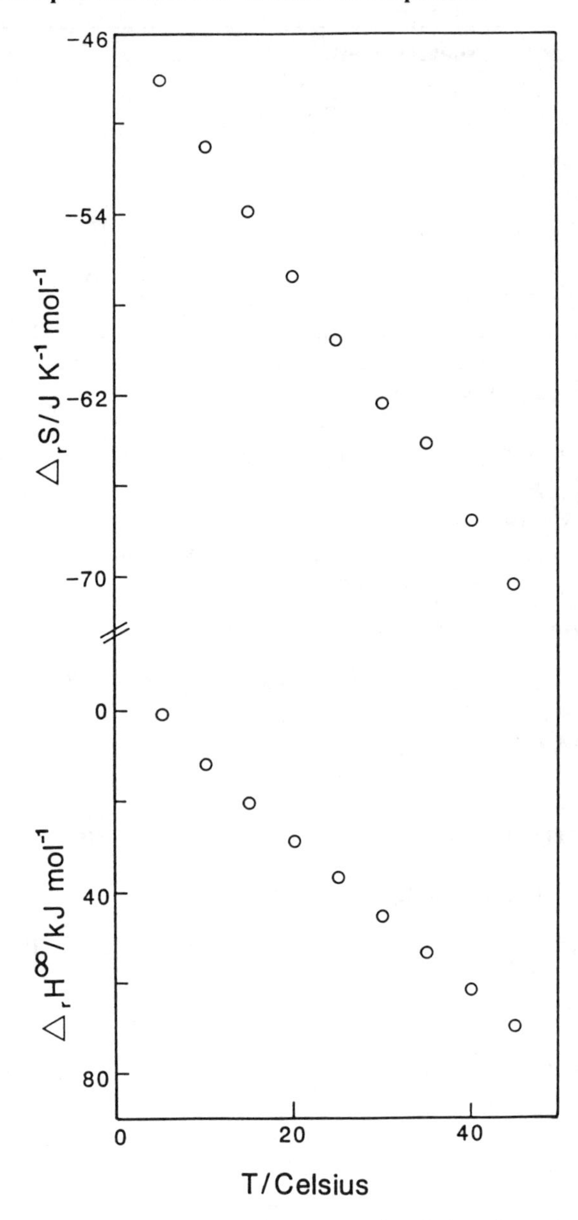

Fig. 6.5.1. Dependence on temperature of entropy and enthalpy of reaction for cyano-ethanoic acid(aq).

constant of 2-cyanoethanoic acid(aq). The several difference parameters are summarized in Table 6.5.2 where we start with the dependence of $\Delta_r G^0(T)$ on temperature. The difference column for ε_3 shows an alternation in sign between the individual differences; the mean difference increases on going to the next set of differences in ε_4. We conclude that no conclusions can be drawn concerning the dependence on temperature of $\Delta_r C_p^\infty$. With this information to hand we embark on the calculation using two linear least squares procedures.

As commented above each difference ε_1 $[=\{\Delta_r G^0(T+\Delta T) - \Delta_r G^0(T)\}/\Delta T]$ is a measure of $\Delta_r S$ at the mean temperature $T + (\Delta T/2)$. If $\Delta_r C_p^\infty$ is independent of temperature, the dependence of $\Delta_r S^0$ on temperature is given by the equation in Table 6.5.3 which uses the reference temperature 3.5 K $\{=1+(\Delta T/2)\}$. A least squares analysis shows that this equations is satisfactory (Table 6.5.4), the plot of $\Delta_r S^0$ against $\ln(T/\mathrm{K})$ being linear. The next stage of the analysis is set out in Table 6.5.5 with the aim of calculating the enthalpy of reaction (Table 6.5.6). The calculated dependence of $\Delta_r H^\infty(T)$ on temperature is shown in Fig. 6.5.1.

References to section 6.5

[1] D. J. G. Ives; P. G. N. Moseley. *J. Chem. Soc. Faraday Trans. 1*, 1976, **72**, 1132.
[2] D. H. Everett; W. F. K. Wynne-Jones. *Trans. Faraday Soc.*, 1939, **35**, 1380.
[3] F. S. Feates; D. J. Ives. *J. Chem. Soc.*, 1956, 2798.
[4] L. E. Strong; T. Kinney; P. Fischer. *J. Soln. Chem.*, 1979, **8**, 329.

7

Dependence of equilibrium composition and rate constants on temperature and pressure

7.1 INTRODUCTION

The treatments described in Chapters 4 to 6 concern the dependences of equilibrium and rate constants on temperature at fixed pressure and on pressure at fixed temperature. We consider in this chapter cases where the data describe the dependence of these constants on both T and p. These dependences of related thermodynamic variables are related by Maxwell equations (Table 7.1.1), relationships based on thermodynamic definitions. In other words the outcome of the analysis of a given set of data must satisfy these equations. No new conclusions [1] can be drawn about the underlying chemistry. If the data do not conform, a mistake has been made.

A possible approach to the analysis of the data is described in Table 7.1.2 in which $K(T;p)$ is fitted about $K(\theta;\pi)$ where θ and π are reference temperatures and pressures respectively. In effect, we expand the dependence of $\ln K$ on T and p about $\ln K(\theta;\pi)$

Table 7.1.1. Maxwell equations

Assuming pressure $p \simeq p^0$.
Equilibrium composition:

$$[\partial \Delta_r H^\infty / \partial p]_T = \Delta_r V^\infty(\text{sln}; T; p) - T[\partial \Delta_r V^\infty(\text{sln})/\partial T]_p$$

$$[\partial \Delta_r S^0(\text{sln})/\partial p]_T = -[\partial \Delta_r V^\infty(\text{sln})/\partial T]_p$$

$$[\partial \Delta_r C_p^\infty(\text{sln})/\partial p]_T = -T[\partial^2 \Delta_r V^\infty(\text{sln})/\partial T^2]_p$$

Kinetics:

$$[\partial \Delta^{\neq} H^\infty / \partial p]_T = \Delta^{\neq} V^\infty(\text{sln}; T; p) - T[\partial \Delta^{\neq} V^\infty(\text{sln})/\partial T]_p$$

$$[\partial \Delta^{\neq} S^0(\text{sln})/\partial p]_T = -[\partial \Delta^{\neq} V^\infty(\text{sln})/\partial T]_p$$

$$[\partial \Delta^{\neq} C_p^\infty(\text{sln})/\partial p]_T = -T[\partial^2 \Delta^{\neq} V^\infty(\text{sln})/\partial T^2]_p$$

Table 7.1.2. Dependence of ln *K* on temperature and pressure

$$\ln K(T;p) = a_1 + a_2(T-\theta)/\mathrm{K} + a_3(p-\pi)/\mathrm{Pa} + a_4[(T-\theta)/\mathrm{K}]^2 + a_5[(T-\theta)(p-\pi)/\mathrm{K\,Pa}] + a_6[(p-\pi)/\mathrm{Pa}]^2 + \ldots$$

$\ln K(\theta;\pi) = a_1$ where θ = reference temperature and π = reference pressure.

Table 7.1.3. Dependence on pressure (six-term equation)

$$(\partial \ln K/\partial p)_T = a_3/\mathrm{Pa} + [a_5(T-\theta)/\mathrm{K\,Pa}] + [2a_6(p-\pi)/\mathrm{Pa}^2]$$

Hence, $\Delta_r V^\infty(T;p) = -RT[a_3/\mathrm{Pa} + a_5(T-\theta)/\mathrm{K\,Pa} + 2a_6(p-\pi)/\mathrm{Pa}^2]$

$$\Delta_r V^\infty(\theta;\pi) = -R\theta a_3/\mathrm{Pa}$$

$(\partial\Delta_r V^\infty/\partial T)_p = -R[a_3/\mathrm{Pa} + a_5(T-\theta)/\mathrm{K\,Pa} + 2a_6(p-\pi)/\mathrm{Pa}^2] - RT[a_5/\mathrm{K\,Pa}]$
At $(\theta;\pi)$, $\mathrm{d}\Delta_r V^\infty/\mathrm{d}T = -R[a_3/\mathrm{Pa} + \theta a_5/\mathrm{K\,Pa}]$
Also, $(\partial\Delta_r V^\infty/\partial p)_T = -2RTa_6/(\mathrm{Pa})^2$

Table 7.1.4. Dependence on temperature

From Table 7.1.2,
$(\partial \ln K/\partial T)_p = a_2/\mathrm{K} + 2a_4(T-\theta)/\mathrm{K}^2 + a_5(p-\pi)/\mathrm{K\,Pa}$
Then, $\Delta_r H^\infty(T;p) = RT^2[a_2/\mathrm{K} + 2a_4(T-\theta)/\mathrm{K}^2 + a_5(p-\pi)/\mathrm{K\,Pa}]$
$\Delta_r H^\infty(\theta;\pi) = R\theta^2 a_2/\mathrm{K}$; $(\partial\Delta_r H^\infty/\partial p)_T = RT^2 a_5/\mathrm{K\,Pa}$
$\Delta_r C_p^\infty = (\partial\Delta_r H^\infty/\partial T)p$
$= 2RT[a_2/\mathrm{K} + 2a_4(T-\theta)/\mathrm{K} + a_5(p-\pi)/\mathrm{K\,Pa}] + RT^2 2a_4/\mathrm{K}^2$
At $(\theta;\pi)$, $\Delta_r C_p^\infty = 2R\theta a_2/\mathrm{K} + 2R\theta^2 a_4/\mathrm{K}^2$
From Table 7.1.1 (Maxwell Equation) at $(\theta;\pi)$
$\Delta_r V^\infty - \theta(\partial\Delta_r V^\infty/\partial T)_p = -R\theta a_3/\mathrm{Pa} + \theta R[a_3/\mathrm{Pa} + \theta a_5/\mathrm{K\,Pa}]$
$= R\theta^2 a_5\mathrm{K}/\mathrm{Pa} = (\partial\Delta_r H^\infty/\partial p)_\theta$

at θ and π. In Tables 7.1.3–7.1.4, we consider a system where the data are fitted satisfactorily using six or less terms in the equation. A similar set of equations are readily derived for the dependence of $k^\#$ on T and p about $k^\#(\theta;\pi)$ at temperature θ and pressure π.

Reference to section 7.1

[1] S. D. Hamann. *Aust. J. Chem.*, 1984, **37**, 867.

7.2 MARSHALL AND FRANCK EQUATION

The Marshall and Franck equation (Table 7.2.1) combines a simple equation for the temperature dependence with the Marshall and Mesmer equation for the pressure dependence (Chapter 4). The resulting equation describes the dissociation of $CaCl_2$

Table 7.2.1. Marshall and Franck equation

$$\ln K(\text{aq}; T; p) = a_1 + a_2(\text{K}/T) + a_3(\text{K}/T)^2 + a_4(\text{K}/T)^3 + [a_5 + a_6(\text{K}/T) + a_7(\text{K}/T)^2] \ln \{\rho_1^*(1; T; p)/\text{kg m}^{-3}\}$$

Then for systems where $\rho_1^*(1; T; p)/\text{kg m}^{-3}$ is held constant at $\rho_1^*(1; \theta; \pi)$,
At constant ρ_1^*,

$$\text{d} \ln K/\text{d}T = -(a_2\text{K}/T^2) - (2a_3\text{K}^2/T^3) + [-(a_6\text{K}/T^2) - (2a_7\text{K}^2/T^3)] \ln\{\rho_1^*(\theta; \pi)/\text{kg m}^{-3}\}$$

and $MgCl_2$ as a function of T and p. In describing the self-dissociation of water over the range $273 < T/\text{K} < 1273$ and $1 < p/\text{bar} < 105$, the seven-term equation was used [3] (Table 7.2.1).

The equations in Table 7.2.1 can be rewritten to express the dependence on temperature of $\ln K$ in systems where the density of the solvent is constant, e.g. unity. If the density is constant, the volume of a given mass of solvent is also constant.

References to section 7.2

[1] J. D. Frantz; W. L. Marshall. *Am. J. Sci.*, 1982, **282**, 1666.
[2] W. L. Marshall. *Pure Appl. Chem.*, 1985, **57**, 283.
[3] W. L. Marshall; E. U. Franck. *J. Phys. Chem. Ref. Data*, 1981, **10**, 295.
[4] R. E. Mesmer; W. L. Marshall; D. A. Palmer; J. M. Simonoson; H. F. Holmes. *J. Soln. Chem.*, 1986, **17**, 699.

7.3 EVANS AND POLANYI; ISOCHORIC QUANTITIES

The treatment described in the previous section prompts an interesting question concerning the dependence of K on temperature at constant density of the solvent. An equivalent condition is with respect to constant molar volume of the solvent, with emphasis on the term 'constant volume'. In general terms, isochoric means 'at constant volume' and denotes a condition imposed on a thermodynamic process in the same way that isothermal and isobaric denote 'at constant temperature' and 'at constant pressure' respectively (Chapter 1).

Nevertheless the isochoric condition is less familiar than the isobaric and isothermal conditions in the context of chemical reactions in solution. In addition there are instances where the term isochoric is used and where the meaning is not quite the same as implied above. Problems emerge from attempts to develop arguments introduced [1] by Evans and Polanyi in the context of understanding activation parameters for reactions in solution [2, 3].

Conventional activation parameters are obtained by analysing the dependences of rate constants on temperature and pressure (e.g. Chapters 4–6). The sign and magnitude of these activation parameters are determined by solvent–solvent and solute–solvent interactions. However, the strengths of these interactions at temperature T differ from the strengths at temperature $T + \delta T$, at fixed pressure. In other words when the temperature changes so does the solvent. Similar comments apply to the dependence of rate constants on pressure at fixed temperature; when the pressure changes so does the solvent. Evans and Polanyi [1] drew attention to these points in their paper published in 1935. The main thrust of their paper was directed towards the calculation of volumes of activation (see Chapter 4). But in dealing with the dependence of rate constants on temperature they noted that interpretation of kinetic data would be simpler if it were possible to keep solute–solvent distances constant. Clearly this is impossible. In order to understand much of what follows it is helpful to quote Evans and Polanyi who wrote the following [1]:

> Especial difficulty arises in solution from the interaction between solvent and solute which depends strongly on temperature. This effect would be *to some extent* eliminated by measuring the temperature coefficients at constant volume instead of the current method of constant pressure.

The three italicized words signal that Evans and Polanyi recognized a compromise [1].

References to section 7.3

[1] M. G. Evans; M. Polanyi. *Trans. Faraday Soc.*, 1935, **31**, 875.

[2] J. B. F. N. Engberts; J. Haak; M. J. Blandamer. *J. Amer. Chem. Soc.*, 1985, **107**, 6031.

[3] M. J. Blandamer; J. Burgess; B. Clark; J. M. W. Scott. *J. Chem. Soc. Faraday Trans. 1*, 1984, **80**, 3359.

[4] M. J. Blandamer; J. Burgess; J. B. F. N. Engberts. *Chem. Soc. Reviews*, 1985, **14**, 237.

7.4 ISOCHORIC

In Chapter 1 we noted how the Gibbs energy G is central to the theme of this text because we know that at fixed T and p, closed systems at equilibrium are at minima in G. In other words G is the thermodynamic potential function for isobaric–isothermal conditions. The Helmholtz energy plays the same role for systems held at constant temperature and constant volume, i.e. isothermal and isochoric conditions. To emphasize the point, here isochoric means the volume of the system is held constant. During a spontaneous process leading to equilibrium the pressure p will change. This use of the term isochoric is thermodynamically correct. However, we now use the term 'isochoric' to mean something slightly different.

A given solution, solvent = liquid 1_l, contains a chemical equilibrium characterized by $K(T;p)$ at temperature T and pressure p. In other words $K(T;p)$ describes the composition of a solution at a minimum in the Gibbs energy G at defined T and p (Chapter 2). At temperature T and pressure p, the molar volume of the solvent, liquid 1_l is defined by the independent variables T and p (Table 7.4.1). We assume that for

Table 7.4.1. Equilibrium and volumetric parameters

K describes a chemical equilibrium; solvent = liquid substance 1.

$\ln K = \ln K[T;p] \quad \mathrm{d}\ln K = (\partial \ln K/\partial T)_p \, \mathrm{d}T + (\partial \ln K/\partial p)_T \, \mathrm{d}p$

$V_1^* = V_1^*[T;p] \quad \mathrm{d}V_1^* = (\partial V_1^*/\partial T)_p \, \mathrm{d}T + (\partial V_1^*/\partial p)_T \, \mathrm{d}p$

a given chemical equilibrium the dependence of $K(\text{sln}; T; p)$ on both T and p has been delineated. Further we assume that the temperature and pressure dependences of the volumetric properties of the solvent (Table 7.4.1) are also available.

Hence at temperature θ and pressure π, $K(\text{sln}; \theta; \pi)$ and $V_1^*(1; \theta; \pi)$ are known together with the corresponding partial derivatives described in Table 7.4.1. We assume that there also exists a pressure $(p + \Delta p_1^*)$ where,

$$V_1^*(1; \theta; \pi) = V_1^*(1; \theta + \delta\theta; \pi + \delta\pi_1^*) \tag{7.4.1}$$

Here $\delta\pi_1^*$ is characteristic of the solvent, temperature and pressure. In other words the molar volumes of the solvent are equal at $(\theta; \pi)$ and $(\theta + \delta\theta; \pi + \delta\pi_1^*)$. Equation (7.4.1) identifies an isochoric condition with respect to the molar volume of the solvent. Hence two equilibrium constants $K(\text{sln}; \theta; \pi)$ and $K(\text{sln}; \theta + \delta\theta; \pi + \delta\pi_1^*)$ can be compared for conditions where the molar volumes of the solvent are equal. The latter isochoric condition is extrinsic to the system characterized by $K(\text{sln}; \theta; \pi)$ and $K(\text{sln}; \theta + \delta\theta; \pi + \delta\pi_1^*)$. The volumes of the corresponding solutions supporting these equilibria are unlikely to be equal at $(\theta; \pi)$ and $(\theta + \delta\theta; \pi + \delta\pi_1^*)$.

We have used the word 'extrinsic' (not inherent nor intrinsic, but extraneous, not belonging) [1] in conjunction with the word isochoric, thereby drawing a clear distinction with the isochoric condition discussed with reference to the Helmholtz function F. In the latter case the isochoric condition is intrinsic to the system.

In general terms therefore, we track from $\ln K(\theta; \pi)$ at given temperature θ and pressure π a dependence of $\ln K$ on T and p under the extrinsic condition that the molar volume of the solvent V_1^* at all T and p equals $V_1^*(\theta; \pi)$.

The major problem in setting out this type of analysis concerns the availability of volumetric data describing the dependence of V_1^* over extensive ranges of T and p. In the case of equilibrium constants for aqueous solutions such data are available. In fact it is possible to extend the analysis to other properties. For example there is sufficient infomation to examine the relative permittivities and viscosities of water as a function of T and p under the condition that V_1^* is constant.

These dependences stimulate new questions bearing in mind that, on average, at constant V_1^* the intermolecular separation is constant.

Reference to section 7.4

[1] *The Pocket Oxford Dictionary*, Clarendon Press, 4th edn., revised, 1946.

7.5 ISOCHORIC ACTIVATION PARAMETERS

The original proposal [1] of Evans and Polanyi was directed at kinetic parameters rather than equilibrium thermodynamic parameters. In fact an intense debate [2–11] about isochoric parameters is directed towards the significance of isochoric kinetic activation parameters. The suggestion made by Evans and Polanyi has attracted considerable interest. The starting point is the dependence of kinetic parameters $k^{\#}$ on T and p, fitted to the general equation in Table 7.5.1. There are advantages from a linear least squares standpoint in using the reduced variables $(T - \theta)/\theta$ and $(p - \pi)/\pi$. Hence the enthalpies and volume of activation can be calculated as a

Table 7.5.1. Kinetic parameters

$$\ln(k^{\#}\,\mathrm{K}/T) = a_1 + a_2(T-\theta)/\theta + a_3(p-\pi)/\pi + a_4[(T-\theta)/\theta]^2 + a_5[(p-\pi)/\pi]^2 + a_6[(p-\pi)/\pi][T-\theta)/\theta] + \ldots$$

Restricting attention to six terms;

$$\{\partial \ln(k^{\#}\,\mathrm{K}/T)/\partial T\}_p = (a_2/\theta) + 2a_4(T-\theta)^2/\theta^2 + a_6(p-\pi)/\theta\pi$$

Then, $\Delta^{\neq} H^{\infty}(\mathrm{sln}; T; p) = RT^2[(a_2/\theta) + 2a_4(T-\theta)^2/\theta^2 + a_6(p-\pi)/\theta\pi]$
and $\Delta^{\neq} H^{\infty}(\mathrm{sln}; \theta; \pi) = R\theta a_2$
$\Delta^{\neq} V^{\infty}(\mathrm{sln}; T; p) = -RT[(a_3/\pi) + (2a_5(p-\pi)/\pi^2) + a_6(T-\theta)/\theta\pi]$
$\Delta^{\neq} V^{\infty}(\mathrm{sln}; \theta; \pi) = -R\theta a_3/\pi$

function of both T and p. In particular we can calculate the activation parameters under conditions $(T;p;)^v$ at which the molar volume of the solvent equals, say, $V_1^*(1; 298.15\,\mathrm{K}; 101\,325\,\mathrm{Pa})$. The outcome is quite thought-provoking. A plot of the dependences on temperature of $\Delta^{\neq} V^{\infty}[\mathrm{s\,ln}; (T;p)^v]$ describes a variation in activation volume under constraint which requires the molar volume of the solvent bathing initial and transition states to be constant.

References to section 7.5

[1] M. G. Evans; M. Polanyi. *Trans. Faraday Soc.*, 1935, **31**, 875.
[2] C. H. Lupis. *Acta Metallurgica*, 1978, **26**, 211.
[3] P. G. Wright. *J. Chem. Soc Faraday Trans. 1*, 1986, **82**, 2557.
[4] E. Whalley. *J. Chem. Soc Faraday Trans. 1*, 1986, **82**, 2557.
[5] M. J. Blandamer; J. Burgess; B. Clark; J. M. W. Scott. *J. Chem. Soc. Faraday Trans. 1*, 1984, **80**, 3359.
[6] L. M. P. C. Albuquerque; J. C. R. Reis. *J. Chem. Soc Faraday Trans. 1*, 1989, **85**, 207.
[7] S. D. Brummer; G. J. Hills. *Trans. Faraday Soc.*, 1961, **57**, 1816, 1823.
[8] M. J. Blandamer; J. Burgess; B. Clark; R. E. Robertson; J. M. W. Scott. *J. Chem. Soc. Faraday Trans. 1*, 1985, **81**, 11.
[9] E. Whalley. *Adv. Phys. Org. Chem.*, 1964, **2**, 93.
[10] E. Whalley. *Ber. Bunsenges physik Chemie.*, 1966, **70**, 958.
[11] M. J. Blandamer; J. Burgess; H. J. C. Cowles; I. M. Horn; J. B. F. N. Engberts; S. A. Galema; C. D. Hubbard. *J. Chem. Soc. Faraday Trans. 1*, 1989, **85**, 3733.

7.6 RELATED PARAMETERS

The analysis described above can be extended in several directions. Caldin suggested [1] a definition which considered the dependence of rate constants on temperature at constant volume of activation. Another possibility uses the internal pressure of the solvent as a variable. The effects of internal pressure on properties of solutes has been discussed including solvatochromism [2]. Equilibrium constants describing conformational equilibria shows a linear dependence [3] on internal pressure at fixed temperature for four solvent systems. The effects have been discussed of internal pressures on rate constants describing the kinetics of combination and disproportionation of ethyl radicals in 11 solvents [4] and for racemization [5] of 1,1′-binaphthyl.

References to section 7.6

[1] E. A. Caldin, in G. J. Hills; C. A. Vianna. *Hydrogen Bonded Systems*, (ed. A. K. Covington; P. Jones), Taylor and Francis, London, 1968, p. 281.
[2] J. Gordon. *J. Phys. Chem.*, 1966, **70**, 2413.
[3] R. J. Ouellette; S. H. Williams. *J. Amer. Chem. Soc.*, 1971, **93**, 466.
[4] A. P. Stefani. *J. Amer. Chem. Soc.*, 1968, **90**, 1694.
[5] A. K. Colter; L. M. Clemens. *J. Phys. Chem.*, 1964, **68**, 651.

Index

activation, 94
  energy, 57
  standard molar enthalpy of, 57
  standard molar entropy of, 57
activity coefficients
  in a liquid mixture, 31
  salt in solution, 39
  solute in solution, 33
  solvent in solution, 31
affinity, 3
  Gibbs energy, 5
  pressure, 10
  spontaneous change, 3
  stability, 13
  temperature, 13
axiom (thermodynamic), 3

binary aqueous mixture of hydrogen ions, 57

chemical equilibrium, 48
chemical potential
  affinity, 20
  Gibbs energy, 21
  Gibbs theorem, 21
  and partial molar enthalpy, 30
  and partial molar entropy, 29
  solute, 35
  solvent, 32
  standard, 30, 32, 35
  thermodynamic energy, 19, 20
compressibility
  isothermal, 10
    equilibrium, 26
    frozen, 26
compression, 17
correlation matrix, 96, 117

description
  Gibbs energy, 22
  system, 23

egg box, 22
energy
  Gibbs energy, 5
    description of system, 25
    standard reaction, 48
  Helmholtz, 8
  thermodynamic, 1, 4
enthalpy, 7
  partial molar, 22
  pressure dependence, 11
  of reaction, 38
entropy, 2
  description of system, 25
  partial molar, 22
  pressure dependence, 11
  temperature dependence, 11
equations
  Arrhenius, 94
  Asona, 82
  Benson–Berson, 72
  Clarke–Glew, 114
  El'yanov and Gonikberg, 77
  El'yanov and Hamann, 79
  El'yanov and Vasylvitskaya, 81
  Feates and Ives, 106
  Gibbs–Duhem, 28, 37
  Gibbs–Helmholtz, 9, 28
  Gurney, 130
  Halliwell–Strong, 125
  Harned and Embree, 87, 108
  Ives and Pryor, 104
  Lown, Thirsk and Wynne-Jones, 76
  Magee, Ri and Eyring, 110
  Marshall and Franck, 137

Marshall and Mesmer, 84
Maxwell's, 9, 138
Nakahara, 82
North, 69
Owen–Brinkley, 66, 75
polynomial, 119
Robinson, 100
Swaddle, 71
Tait, 17, 69, 72, 75, 77
Valentiner, 90
van't Hoff, 50, 88
equation of state, 16
equilibrium constant, 48
pressure dependence, 49, 50, 60
linear dependence, 63
quadratic dependence, 65
standard, 48, 50
temperature dependence, 49, 50
linear dependence, 89
expansivity, 10
extrathermodynamic assumption, 79

first law, 2

gas
chemical potential, 30
ideal gas, 30

heat, 1
heat capacity
equilibrium, 7, 12
frozen, 7, 12
isobaric, 7
partial molar, 22
standard reaction, 48
hydrogen ions, 43
chemical equilibria, 52
chemical potential, 44

internal pressure, 16, 141
isochoric, 138
activation, 140

kinetics of chemical reaction, 51

laws (of thermodynamics)
first law, 2
second law, 2
Le Chatelier's principle, 15
liquid mixtures
chemical potentials, 31, 32
enthalpies, 31
volumes, 31

orthogonal polynomials, 123

partial molar
enthalpy, 34, 37
entropy, 37
heat capacity (isobaric), 34
volume, 34
practical osmotic coefficient, 32
pressure
gauge, 68
reference, 63
standard, 30

rate constant
first-order, 52
pressure dependence, 61
second-order, 53
temperature dependence, 86
linear, 89
reference temperature, 87

salt, 38
chemical potential, 39
mean ionic activity coefficient, 39, 46
partial molar
enthalpy, 41
entropy, 42
volume, 41
sigma method, 131
solute, chemical potential, 32, 33
solvates, 42
solvatochromism, 141

Taylor expansion, 62, 110
temperature functions, 93
transmission coefficient, 57
transition state theory, 51

van't Hoff theorem, 15
variables
dependent and independent, 1
volume
description of system, 26
of activation, 61
of reaction, 38, 63
partial molar, 22

water
self-dissociation, 57
work, 1